International Geological Congress

Guide to Washington and its Scientific Institutions

International Geological Congress

Guide to Washington and its Scientific Institutions

ISBN/EAN: 9783337417482

Printed in Europe, USA, Canada, Australia, Japan

Cover: Foto ©berggeist007 / pixelio.de

More available books at **www.hansebooks.com**

GUIDE TO
WASHINGTON

AND ITS

SCIENTIFIC

WITH
TWO MAPS.

INSTITUTIONS

INTERNATIONAL CONGRESS OF GEOLOGISTS.

FIFTH SESSION. WASHINGTON.

1891

OPENING SESSION, WEDNESDAY, AUGUST 26, 2 P. M.
COLUMBIAN UNIVERSITY.

Prepared by the Local Committee for the use of the Congress.

LOCAL COMMITTEE.

GARDINER G. HUBBARD, *Chairman.*

CHARLES D. WALCOTT, *Secretary.*

MARCUS BAKER, S. F. EMMONS, PROF. S. P. LANGLEY,
DR. G. F. BECKER, G. K. GILBERT, W. J. MCGEE,
WHITMAN CROSS, G. BROWN GOODE, DR. T. C. MENDENHALL,
DR. W. H. DALL, ARNOLD HAGUE, MAJ. J. W. POWELL,
DR. DAVID T. DAY, JOS. P. IDDINGS, HON. EDWIN WILLETTS,
MAJ. C. E. DUTTON, U. S. A. BAILEY WILLIS.

EXECUTIVE COMMITTEE.

GARDINER G. HUBBARD, *Chairman.*

DR. W. H. DALL, G. K. GILBERT, MAJ. J. W. POWELL,
S. F. EMMONS, ARNOLD HAGUE, CHARLES D. WALCOTT.

COMMITTEE ON HALLS AND PLACE OF MEETING.

DR. W. H. DALL, *Chairman.*

G. BROWN GOODE, W. J. MCGEE.

COMMITTEE ON PRINTING.

MAJ. J. W. POWELL, *Chairman.*

DR. T. C. MENDENHALL. HON. EDWIN WILLETTS.

COMMITTEE ON SHORT EXCURSIONS.

G. K. GILBERT, *Chairman.*

MAJ. C. E. DUTTON, U. S. A. BAILEY WILLIS.

COMMITTEE ON FINANCE.

GARDINER G. HUBBARD, *Chairman.*

MARCUS BAKER, JOS. P. IDDINGS.

COMMITTEE ON ENTERTAINMENT.

ARNOLD HAGUE, *Chairman.*

WHITMAN CROSS, JOS. P. IDDINGS,
DR. DAVID T. DAY, BAILEY WILLIS.

The District of Columbia.

THE District of Columbia is the permanent seat of government of the United States, and Washington is the capital city within its bounds. Its exact site was chosen by President Washington in accordance with a resolution passed by Congress July 10, 1790, which specified that the location should be upon the banks of the Potomac river between certain limits. This choice was reached after a heated sectional contest, and conformed to the declaration that "the site of the future capital should be as near as possible the centre of wealth, of population, and of territory."

The District was originally ten miles square ; its centre was very near the spot occupied by the Washington Monument, and jurisdiction was ceded to the general government by the States of Maryland and Virginia. The boundaries of the District as originally defined are shown upon the accompanying geological map. When the site was chosen Georgetown had been a thriving trading point, with extensive foreign commerce, for nearly a hundred years, and Alexandria was also a prominent settlement, but the ground occupied by the present City of Washington was for the most part unimproved.

The City of Washington.

THE ground plan of the capital is the work of Major L'Enfant, a young French engineer residing in Philadelphia, chosen by Washington for this purpose. The plan was made after a careful study of the physiography of the District, and shows a wonderful appreciation of the requirements of the capital of a great nation.

The Capitol is the centre of this plan. The north and south and east and west lines passing through that building divide the city into four quarters. Either side from the meridian line the streets are numbered : First street, Second street, etc. Each way from the east-west line the streets are named in order from the alphabet : A street, B street, etc. Besides the lettered and numbered streets there are many avenues, named after states of the Union. These avenues run in directions diagonal to the streets and are so arranged that several of them intersect at certain important points— as at the White House and at the Capitol.

3

The streets and avenues of the city are so wide (80 to 160 feet) that in most cases only the central part is used for pavement and sidewalks, leaving a strip on either side which holders of adjoining property are allowed to improve with flowers, shrubs and trees, but may not encroach upon with buildings.

The largest park within the city limits is that known as the Mall, which lies between the Capitol and the Washington Monument. In various divisions of this park are situated the Smithsonian Institution, the National Museum, and other scientific bureaus and museums. (See map). Between the White House grounds and the Monument is the President's Park, commonly known as the "White Lot."

Within a few years the park area of the city will be more than doubled by the improvement of the grounds now being reclaimed, by dredging and filling, from the malarial flats of the river. This land adjoins the Mall on the west and extends southward to a point opposite the Arsenal grounds.

One of the most beautiful features of the city is the great number of small parks, most of them situated at points of intersection of several avenues, while in other cases one or more squares are thus occupied. Perhaps the most beautiful of these small parks is Lafayette Square, situated in front of the White House, between Pennsylvania avenue and H street, and surrounded by houses with which many events of historic interest are connected. It contains a great variety of beautiful trees, many of them exotics. In the centre of this park is an equestrian statue of Andrew Jackson, seventh President of the United States. At the southeastern corner of the park is the monument recently erected to the memory of Lafayette and his compatriots, Count de Rochambeau and Chevalier Duportail, of the French army, and Counts D'Estaing and De Grasse, of the French navy, who served as allies in the closing years of the Revolutionary war. The statue, which was ordered by Congress at a cost of $50,000, was designed by the French artists, Falquière and Mercie.

On Vermont avenue are three pretty parks. McPherson Square, situated between I and K streets, contains an equestrian statue to General J. B. McPherson, erected by the Society of the Army of the Tennessee. Two blocks further up Vermont avenue, at the intersection of Massachusetts avenue, is Thomas Circle, in the centre of which is a statue of General George H. Thomas, erected by the Society of the Army of the Cumberland. Two blocks still further out Vermont avenue is Iowa Circle.

On Connecticut avenue, which leads off in a northwesterly direction from Lafayette Square, is Farragut Square, between I and K streets. In this is a statue of Admiral David G. Farragut. Four blocks up the avenue is Dupont Circle, in the centre of which is a statue to Rear Admiral Samuel F. Dupont.

In Scott Circle, at the intersection of Sixteenth street and Massachusetts avenue, stands an equestrian statue of General Winfield Scott.

On East Capitol street, in the eastern section of the city, is the fine Lincoln Park, with a statue representing the emancipation of the slave.

4

Besides the parks above mentioned the visitor will find many others, at short intervals, on nearly all the principal avenues of the city. The grounds about the Naval Observatory, the Arsenal, and at the Congressional Cemetery are also improved as parks.

The Botanical Gardens are situated on Pennsylvania avenue, between First and Third streets. They cover ten acres of ground, and are beautifully laid out with trees, shrubs and flowers. They may be considered as forming a part of the Mall, although enclosed by an iron railing. Admission may be had between 9 a. m. and 6 p. m. every day except Sunday. The grounds and greenhouses are well worth a visit.

The Zoological Park.

The grounds of the new Zoological Park in the near suburbs of the city lie on both sides of Rock Creek, just north of Woodley lane. They comprise 166 acres. The land was purchased by Act of Congress in 1889 at a cost of nearly $200,000. Already the park is enclosed, and several structures suitable for the use of the animals have been erected. A considerable number of North American animals have been placed in their new home, and an excellent nucleus started for a national zoological garden. A number of the larger Rocky Mountain animals have been captured in the Yellowstone National Park, and will be transferred to Washington for the Zoological Park early in the autumn.

The park is most picturesquely located and admirably adapted for its purpose. It is under the direction of the Smithsonian Institution.

The Capitol.

The Capitol as it now stands is the result of several additions to and changes of the original building. The central part, exclusive of the dome, represents the original design by Mr. Stephen Hallet. The two wings of this part, erected in 1793–1811, were destroyed by the British in 1814. but were soon rebuilt with the connecting portion and a wooden dome. The extensions on the north and south, containing the present legislative chambers, were added 1851–1867, after the plans of Mr. Thomas U. Walter, and the great iron dome, by the same architect, was completed in 1863.

The length of the building is 751 feet, its greatest width 320 feet, and the dome rises 307 feet above the foundation.

In the different facades of the Capitol are 134 beautiful Corinthian columns, 100 of them monolithic. The material of the new wings is white marble, that of the older part sandstone.

At the eastern front of the building, flanked by a double row of columns, is a portico 160 feet long, upon which most of the Presidents have been inaugurated.

The Capitol contains the legislative chambers of the Senate and of the House of Representatives, the United States Supreme Court Room and Congressional Library.

The rotunda of the Capitol is 96 feet in diameter at its base, and 185 feet high, to a canopy 65 feet in diameter. In the rotunda are eight large paintings by American artists, four of them commemorating events in the discovery and settlement of the country, and four representing scenes in the Revolutionary war. The frieze, 10 feet in height, is likewise historical in character. In the canopy is an allegorical fresco, the apotheosis of Washington, by Brumidi, who also began the frieze.

From the rotunda one can ascend to the dome and to the cupola above, from which a beautiful view of the city may be obtained. The dome is crowned by a bronze statue of the Goddess of Freedom, by Crawford, an American sculptor. The dome is 135 feet 5 inches in diameter at its base.

At the entrance to the rotunda from the eastern portico is a bronze door representing in its relief figures the history of Columbus and his discoveries. There are also heads of many sovereigns and discoverers whose names are associated with the discovery of America, and of historians who have written upon the subject. The door was designed by Randolph Rogers, in 1858. Another fine bronze door is at the eastern entrance to the Senate wing. This was designed by Crawford, and was cast at Chicopee, Massachusetts.

The assembly halls of the Senate and of the House of Representatives, and the rooms connected with them, are ornamented with many frescoes, paintings, and artistic decorations. Attention is especially called to two large paintings by Thomas Moran, situated in the vestibule to the ladies' gallery of the Senate chamber. One of these represents the "Grand Cañon of the Yellowstone," and the other the "Grand Cañon of the Colorado." Both are well worthy of study. Among the miscellaneous paintings which adorn the halls and galleries are "Westward Ho," by Leutze, and the "Signing of the Declaration of Independence."

Between the rotunda and the House wing of the building is the National Hall of Statuary. To this collection each State of the Union has been invited to contribute two statues of prominent citizens. Many of them have already done so.

The White House.

THE Executive Mansion, or White House, is situated in a park between the Treasury and the State, War and Navy buildings. It was erected in 1792–1799, after the designs of Mr. James Hoban, and is said to be similar to the palace of the Duke of

Leinster, in Dublin. Its popular name is said to have its origin in the fact that for a long time after its completion it was the only white building in the city.

The largest of the reception rooms is open to visitors from 10 a. m to 3 p. m. Concerts by the Marine Band are given at 6 o'clock every Saturday afternoon during the summer in the grounds south of the White House.

The Washington Monument.

THE Washington Monument stands on the bank of the Potomac river south of the White House, very near the spot designated by Major L'Enfant in the original plan of the city for an equestrian statue to the memory of Washington. It is also very near the centre of the original District of Columbia.

The designer of the Monument was Robert Mills, of South Carolina. Its erection was begun in 1847, but was interrupted in 1855, when it had reached a height of 152 feet, through failure of funds, which had thus far been contributed by private individuals. Work was resumed in 1878 under appropriations made by Congress. The capstone was put in place December 6, 1884, and the dedication took place on February 21, 1885, with imposing Masonic ceremonies. Robert C. Winthrop, of Massachusetts, was the orator both at the laying of the cornerstone and at the dedication. The total cost of the Monument has been $1,200,000, of which $300,000 was raised by contributions from the people.

The shaft is of Vermont marble. Its original foundation was 80 feet square at the base, 55 feet square at the top and 25 feet high, 17 feet above the surface. When work was resumed in 1878 it was found advisable to enlarge the foundation, and a mass of concrete 126½ feet square and 13½ feet in thickness was placed under the original foundation, a noteworthy feat of engineering. The engineer in charge of the work from 1878 to the completion of the Monument was Col. (now Gen.) Thomas L. Casey.

The Monument is 555 feet in height, 55 feet square at the base, and 31½ feet square at the base of the summit pyramid, which is 55 feet high. The apex of the pyramid is a solid block of aluminum 9 inches high, 4½ inches square at the base and weighing 6¼ pounds. The total weight of the Monument is 80,000 tons. At the time of its completion this shaft was the highest building in the world. It is now (1891) surpassed only by the Eiffel Tower in Paris.

By means of an elevator one can ascend to a landing at the base of the summit pyramid, and through port holes obtain magnificent views of the city and surrounding country. By walking down the iron staircase one can see the numerous memorial tablets set in the walls, contributed by various nations, states, cities, societies, corporations and individuals.

7

The elevator ascends at the even hour and half hour. The Monument is open every week day from 9 a. m. to 5.30 p. m.

The Corcoran Art Gallery.

SITUATED on Pennsylvania avenue, corner of Seventeenth street, opposite the State, War and Navy Departments. This Gallery was founded and endowed by W. W. Corcoran, a citizen of Washington. The present building was erected in 1859. The two bronze lions at the main entrance are copies of Cantora's at the tomb of Pope Clement XIII. It has one of the best collections of paintings in this country, and is constantly being enriched by purchase. Connected with the Gallery is a school of art. Unfortunately the Gallery is closed for repairs during August.

The Departments and Scientific Institutions.

Building of the State, War and Navy Departments.

THIS massive structure stands on the south side of Pennsylvania avenue just west of the White House. It is built in Italian Renaissance style, and was begun in 1871 and completed in 1887, from designs by Mr. A. B. Mullett, late supervising architect of the Treasury. The stone is granite, from Maine and Virginia. The State Department occupies the southern portion of the building ; the War Department the northern and western, and the Navy Department the eastern wing. Many of the rooms are richly frescoed and decorated, and contain numerous portraits, historical relics and other objects of interest.

State Department.

Honorable JAMES G. BLAINE, Secretary of State.

The Department is open from 9 a. m. to 2 p. m. On the third floor is an excellent library for the purposes of the Department. The original Declaration of Independence is exhibited in the library with other historical documents, many of them relating to the early days of the country.

War Department.

Honorable REDFIELD PROCTOR, Secretary of War.

Many of the rooms and corridors are adorned with portraits of distinguished generals, most of which may be seen by applying to the messenger at the Secretary's door.

Headquarters of the Army. Major-General John M. Schofield, Commanding. The office is located in the north wing at the east end of the corridor.

Corps of Engineers. Brigadier-General Thomas L. Casey, Chief of Engineers. The Corps of Engineers are charged with all duties relating to fortifications ; with torpedoes for coast defenses ; with all military bridges ; and such services as may be required for these objects. It is also charged with the harbor and river improvements.

9

Ordnance Bureau. Brigadier-General D. W. Flagler. Chief of Ordnance. The Bureau of Ordnance has charge of all the national armories, gun factories, arsenals and ordnance depots, and is expending large sums of money in the manufacture of modern guns.

The Army Medical Museum and Library.

THE Army Medical Museum occupies a portion of the new building erected at the northwest corner of Seventh and B streets southwest, east of the National Museum. The rest of the building is occupied by the Library of the Surgeon-General's Office, a portion of the Record and Pension Division of the War Department, and the Laboratory. The Museum was removed in 1887 from the building formerly known as Ford's Theater (Nos. 509 and 511 Tenth street northwest.)

The Museum was founded and a large portion of the medical and surgical specimens collected during the war of the rebellion. Since the close of the war, however, the officers in charge have continued to collect specimens from the medical officers of the army at the several military posts, and a number of valuable specimens have been contributed by physicians engaged in private practice.

At the close of the fiscal year terminating June 30, 1891, the Museum contained about 10,135 pathological specimens, 3,314 anatomical specimens, 11,500 microscopical specimens, and 1,717 specimens of comparative anatomy. It was visited last year by more than 42,000 persons.

This collection is richer in specimens illustrative of the results of gun-shot wounds, and of the surgical operations which they necessitate than any other collection in the world. In other departments, though it does not equal some of the wealthy and long established museums of Europe, its collections are, nevertheless, by far the most important in America, and are annually increasing in extent and value.

The Library of the Surgeon-General's Office, which occupies a portion of the same building, is the largest and most valuable medical library in the world. At the close of the fiscal year terminating June 30, 1891, it contained about 100,000 books and 150,000 pamphlets, and the number is steadily increasing. Medical men from any part of the country desirous of consulting the works in this library are courteously welcomed and granted free access.

Both the Museum and Library are open to visitors daily, except Sundays, from 9 a. m. to 4 p. m. The Seventh-street Cable Road, which connects with the principal street railroads, carries visitors direct to the Museum.

Navy Department.

Honorable BENJAMIN F. TRACY, Secretary of the Navy.

The Chiefs of the Bureaus of the Navy Department are officers of the United States Navy and part of the Naval establishment. Upon the walls of the Secretary's office are hung some excellent portraits of former secretaries ; in the corridors are to be seen some fine models of the new cruisers. The Library is on the fourth floor.

Naval Observatory.

Captain S. V. McNAIR, U. S. N., Superintendent.

The Observatory is situated on the corner of Twenty-fourth and D streets North-west. It was established in 1842, its object being to promote the ends of navigation. The Observatory is equipped with a 26-inch equatorial mural circle and transit and a prime transit for declinations, and many other notable instruments. Astronomical observations are made in order to establish and correct the data used by the navigator, and all the instruments connected with navigation are tested in this office. Connected with the Observatory is a corps of astronomers of national reputation. The results of the investigations are published annually under the title of "Washington Observations." The Observatory is open to the public on all work-days from 9 a. m. to 4 p. m. A new observatory is being built one mile north of Georgetown, but it is not yet ready for occupancy. It has an excellent position, admirably chosen for its purposes. The grounds surrounding the building embraces about 60 acres.

Nautical Almanac.

Prof. SIMON NEWCOMB, U. S. N., Superintendent.

The Nautical Almanac Office is situated at the northwest corner of Pennsylvania avenue and Nineteenth street ; entrance, No. 810 Nineteenth street. A regular staff of ten assistants is employed in this office.

Annual Publications :
{ The American Ephemeris and Nautical Almanac.
The American Nautical Almanac.
The Atlantic Coasters Nautical Almanac.
The Pacific Coasters Nautical Almanac.

Publications issued at
irregular intervals : { Astronomical Papers of The American Ephemeris.

U. S. Hydrographic Office.

Lieutenant-Commander RICHARDSON CLOVER, U. S. N., Hydrographer.

A branch of the Bureau of Navigation, Navy Department. Offices in the Department building, basement, east front.

Work consists essentially in the supplying to vessels of war and the merchant-marine of charts, sailing directions, light lists, publications relating to marine meteorology, and other information. The object of the office is to secure the earliest possible reliable information from all sources and to put it promptly before those especially interested in navigation.

Branch offices are established in nine of the principal ports of the United States ; each of these is in the charge of a naval officer, with one or more assistants. In this way information is readily collected and promptly circulated.

The Office is divided into the following divisions :

First.—*Chart Construction.* In charge of the actual engraving of charts. Here can be seen every step in the process, from the time the working sheets are received from the surveying vessels until the final chart is printed from the copperplate. About 60 new nautical chart-plates are produced every year, and about 30,000 charts are printed from copperplates.

Second.—*Issue and Supply.* In charge of the issuing and supplying of charts to naval and other vessels. A supply of every chart likely to be required is kept on hand. Including lithographed charts, the office itself issues 863 different charts, about 10,000 copies being sold per year and 7,000 issued to U. S. Naval vessels.

Third.—*Sailing Directions.* This division has general charge of the archives of the office (where all original data are kept, copies of every chart ever issued by any office and now in actual use, and a copy of every chart ever issued by the Hydrographic Office) ; the preparation and publication of sailing directions for various oceans ; the publication and correction of the six volumes of light lists (lists of light-houses) ; and the weekly Notices to Marines, a pamphlet containing mention of all corrections and changes in charts and other publications (circulation about 1,000 copies per week, not counting the reprints of various paragraphs).

Fourth.—*Marine Meteorology.* In charge of the general subject of climate, weather, storms, currents, best sailing and steam routes, etc. The monthly Pilot Chart of the North Atlantic Ocean, the weekly Hydrographic Bulletin, and occasional treatises on storms of various oceans are prepared and published by this division, which has a corps of about 1,000 voluntary observers who take daily observations and send in their reports from every port. The Pilot Chart has a monthly circulation of 3,300 copies, and is supplied free to the voluntary observers in return for their observations. It contains a forecast for the month succeeding the day of issue and a review of the preceding month, showing graphically the direction and force of prevailing winds, the tracks of storms, positions and tracks of derelict vessels, ice, buoys, and other obstructions to navigation.

Fifth.—*Mailing Division.* This has charge of the correspondence with the branch offices and the mailing of all publications.

U. S. Navy Yard.

Commodore J. S. SKERRETT, U. S. N., Commandant.

The Navy Yard is situated on the Anacostia river, southeast of the Capitol. It is reached by the Washington and Georgetown Railroad in cars marked "Navy Yard"; time from Lafayette Square to the Navy Yard, about 35 minutes. It was formerly a ship-yard and many famous vessels were built there. It is now entirely devoted to the construction of modern ordnance, and its various shops are amply equipped with the best modern machinery for the manufacture of large guns. There is a museum of interesting articles in the Yard. The Navy Yard is open to visitors from 7 a. m. until sundown.

U. S. Marine Barracks.

THE Marine Barracks is the long row of buildings on the ground facing Eighth street, two squares north of the Navy Yard. In the armory on the south side are found some interesting old relics.

Treasury Department.

Honorable CHARLES FOSTER, Secretary.

The Treasury Department stands on Fifteenth street, east of the White House. This building, of Grecian Ionic style of architecture, is, like the Capitol, the result of extensions of the original plan. Mr. Thomas U. Walter was in both cases the architect of the extensions, and produced a very harmonious effect. The old part of the building fronts on Fifteenth street, while the extensions form the northern, western and southern fronts. The original portion of the building is of Virginia sandstone, while the stone employed in the extensions is granite from Dix Island, Maine.

Any one visiting the Treasury should not fail to examine the columns of the new portions, as they are monoliths, 31 feet high and nearly 4 feet in diameter. The main objects of interest are the United States Treasury or Cash Room, the Vaults, and the Secret Service Bureau. The Cash Room is ornamented with beautiful marbles from various places. Open to visitors from 9 a. m. to 2 p. m. A guide is sent with visitors to all places open to the public.

United States Mint.

Dr. E. O. Leech, Director.

The Office of the Director of the Mint is in the Treasury Building. The Director has general supervision of all mints and assay offices, the purchase of silver bullion, and the allotment of its coinage. Two annual reports are published, one upon the operations of the mints and assay offices, and a second upon the statistics of the production of the precious metals in the United States. The report for the calendar year 1890 bears the date of. February 26, 1891.

United States Coast and Geodetic Survey and Office of Standard Weights and Measures.

Dr. T. C. Mendenhall, Superintendent.

The Coast and Geodetic Survey is a bureau under the Treasury Department. Its work, begun in 1817, was almost immediately stopped by legislation, but was resumed in 1832 under the direction of Hassler, its first superintendent. He was succeeded by Bache, under whom the Survey reached a fuller development on the plans proposed by his predecessor.

Its objects are primarily to make surveys of the coast and the adjacent waters, and to collocate these surveys by extended trigonometric operations along the coasts and across the interior. It is also charged by law with the duty of furnishing trigonometric points to the several States.

The extent of the surveyed and unsurveyed shore line is estimated at about 145,000 kilometers.

In addition to its mensurational work, which is of the highest degree of precision, the Survey conducts pendulum observations, tidal researches and a general magnetic survey of the whole territory of the United States. The office of the U. S. Standard of Weights and Measures is also under the direction of the Superintendent, and furnishes standards to the several States and verifies weights and measures.

The publications of the Survey are :

Annual Reports, showing progress and containing professional papers.

Charts on various scales, covering the coast line, for the use of navigators.

Coast Pilot, a series of volumes giving minute descriptions of the coast, with sailing directions.

Tide Tables, giving the predicted tides at the chief ports of the United States.

Professional and scientific papers, published separately from the annual reports, but also contained in them.

Bulletins, giving early results of work accomplished.

Notices to Mariners, giving new data in regard to published charts.

The Charts, Tide Tables and Coast Pilot can be purchased at the Coast and Geodetic Survey Office, or at agencies existing in the principal seaport towns, at about the cost of paper and printing.

The other publications are for gratuitous distribution.

The office is located on New Jersey avenue, near B street southeast, just south of the Capitol.

Bureau of Engraving and Printing.

WILLIAM M. MEREDITH, Chief of Bureau.

This Bureau occupies a large brick building situated on the corner of Fourteenth and B streets southwest, a short distance from the Department of Agriculture. Here are engraved and printed all the United States bonds, the paper money of the Government, and the internal revenue stamps. It is regarded as one of the most interesting bureaus to the general visitor. A competent guide is furnished upon application to the Superintendent of the building. Open to visitors from 9 a. m. to 2 p. m.

Department of the Interior.

Honorable JOHN W. NOBLE, Secretary.

This department building occupies the block bounded by F and G and Seventh and Ninth streets northwest, with the main entrance on F street. It is a massive white structure of imposing appearance; the centre is built of sandstone and the wings of white marble, resting upon a basement of granite. Under this department are gathered a large number of bureaus: the Patent Office, the Pension Office, General Land Office, Office of Indian Affairs, Bureau of Education, Commissioner of Railroads, U. S. Geological Survey, and U. S. Census.

United States Geological Survey.

Major J. W. POWELL, Director.

The Geological Survey is a bureau of the Department of the Interior. It was established by Act of Congress, March 3, 1879, the objects as provided for in the Act being the "classification of public lands and examination of the geological structure, mineral resources and products of the National domain." The President appointed Hon. Clarence King as first Director of the Survey. In March, 1881, Mr. King retired from the directorship and was succeeded by Maj. J. W. Powell, under whose guidance the work of the Survey has developed to its present large proportions.

On account of the extent and diversity of its operations this work is at present carried on by a number of coördinate divisions embracing nearly every department of geology and paleontology, with which are associated laboratories for the investigation of chemical and physical problems directly related to geology. The preparation of a topographical map, to serve as a basis upon which the geological features of the country are finally to be laid down, is carried on in the Division of Geography, with which is connected a large force of topographical engineers and a corps of expert lithographers. There is a Division of Mining Statistics and Technology engaged in preparing annual reports, showing for each calendar year the mineral products of the country. There is also a Division of Illustration, with which is connected a complete photographic laboratory for the reproduction of negatives taken in the field, and copying maps and drawings. The Geological Survey Library contains nearly 30,000 volumes, 42.000 pamphlets, and over 22,000 maps. The distribution of the Survey publications is in charge of the Librarian.

The office of the Geological Survey is located in the Hooe Building, No. 1330 F street northwest, where the greater part of the geological and topographical work is elaborated, the field explorations being conducted during the season in all portions of the United States. The paleontological collections and workshops are located either in the Smithsonian Institution or in the U. S. National Museum ; in the latter are also the chemical and physical laboratories. There are branch offices and laboratories of the Survey in various portions of the country, where special work is being carried on by persons connected with universities and colleges. These form a very considerable portion of the scientific force.

The publications of the Survey are :

Annual Reports. By the Director to the Secretary of the Interior, presenting a summary of the plans and operations of the Survey, accompanied by short administrative reports from chiefs of divisions, followed by a number of scientific papers of general interest.

Monographs. Quarto volumes, containing the more important and elaborate publications of the Survey. Seventeen monographs have been published.

Bulletins. Each of these contains but one paper and is complete in itself. They are, for the most part, short articles giving the more important results of an investigation, and do not properly come under the head of Annual Reports or Monographs. Seventy-nine bulletins have been published.

Annual Reports upon the Mineral Resources of the United States.

The Annual Reports are for gratuitous distribution. Monographs and Bulletins are sold at about the cost of publication. A limited number of the Mineral Resources are for gratuitous distribution.

For a detailed account of the general plan and scope of the Survey and its methods of work, see the Eighth Annual Report of the Director for the year 1886–87.

U. S. Patent Office.

Honorable W. E. Simonds, Commissioner.

The Patent Office was organized in its present form in 1836. It occupies certain portions of the main building on F street. As an object of interest to visitors its principal features are the simple massive architecture of the building itself, and the Model Room in the top story, where models of all patented inventions capable of being thus represented are arranged in cases, classified by subjects. The organization includes an Examining Corps with thirty-two divisions, the last two having been added recently on account of the great expansion of the work ; the Issue and Gazette, Drafting, Assignment, or copying divisions, and the Scientific Library. This library may be of somewhat especial interest to scientific men. It aims to embody, as far as conditions admit, the whole literature of human industry, according to its main purpose of assistance to the examiners in their researches. It is a repository of applied, rather than of pure science. It contains about 50,000 volumes, including pamphlets, and is much used by the patent profession and by branches of the Government doing scientific work.

U. S. Bureau of Education.

William T. Harris, LL. D., Commissioner.

This Bureau is situated at the northwest corner of Eighth and G streets northwest. Its functions will be best understood when it is remembered that the Federal Government of the United States does not support or control the schools and colleges of the country. Each State has full jurisdiction over the subject of education, and the public schools are State institutions, subject entirely to State laws. The Bureau of Education is an agency with the especial function of increasing the enlightened directive power of the people with regard to their schools. This function is performed by the publication of annual and special reports, and occasional bulletins and circulars of information upon educational questions.

The material for these reports is collected by extensive correspondence with the officials in charge of State, city and county public school systems, with the presidents and principals of universities, colleges, seminaries, high schools, and other secondary schools, and with the ministers of education of foreign countries and officers and professors of foreign institutions of learning.

The Library of the Bureau contains 17,500 bound volumes, including all important pedagogical works, and 100,000 pamphlets.

U. S. Census.

Honorable ROBERT P. PORTER, Superintendent.

The Census Office is established by act of Congress every ten years. During its short term it employs thousands of clerks, besides enumerators and special agents in all parts of the United States. The executive office is at the corner of Third and G streets northwest. The count of the population for the year 1890 was made at the Inter-Ocean Building on Ninth street, between E and F streets northwest. In this work the ingenious electric counting machines invented by Dr. Hollerith were used and may be seen in operation, together with the electric classifying system. The results thus far published are in the form of bulletins, eighty-four of which have been issued. Copies of most of these can be obtained by application at the executive office.

Pension Bureau.

Honorable GREEN B. RAUM, Commissioner.

The administration of the enormous business of the Pension Office requires a large building. It stands by itself in Judiciary Square, between Fourth and Fifth and F and G streets northwest. It is an imposing edifice, constructed entirely of red brick ornamented with terra cotta. The inauguration balls of March 4, 1885, and March 4, 1889, were given in the central hall.

Department of Agriculture.

Honorable J. M. RUSK, Secretary.

(Established by an Act of Congress, February 9, 1889).

The Secretary of Agriculture is charged with the supervision of all public business relating to the agricultural industry of the country. He exercises advisory supervision over the agricultural experiment stations deriving support from the National Treasury, and has control of the quarantine stations for imported and domestic cattle.

The Assistant Secretary has general control and direction of a large number of scientific divisions in charge of specialists, whose duties may be concisely expressed as follows :

The Statistician collects all information as to the principal crops and farm animals, and obtains similar information from European countries. He publishes a monthly bulletin of the statistics of the agricultural production, distribution and consumption.

The Entomologist obtains and disseminates information regarding insects, and appropriate remedies for their extirpation.

18

The Botanist investigates plants and grasses of agricultural value or of injurious character, and answers inquiries relating to the same, and has charge of the Herbarium.

The Chemist makes analyses of natural fertilizers, vegetable products and other materials which pertain to the interests of agriculture.

The Ornithologist investigates the economic relations of birds and mammals, and recommends measures for the preservation of those species beneficial to crops and the destruction of injurious species.

The Director of the Office of Experiment Stations secures, as far as practicable, uniformity of methods in the work of the stations throughout the country. He also compiles and publishes such of the results of the station experiments as may be deemed necessary.

The Chief of the Bureau of Animal Industry investigates the existence of dangerous contagious diseases of live stock, superintends the measures for their extirpation, and makes original investigations as to the nature and prevention of such diseases ; has charge of the quarantine stations for cattle, and reports on the animal industries of the country.

The Pomologist collects and distributes information in regard to the fruit industry of the United States and the best means for its improvement.

The Chief of the Division of Vegetable Pathology investigates the diseases of plants, and seeks to determine remedies for their mitigation and prevention.

The Chief of the Division of Forestry is occupied with experiments and reports regarding forestry ; with the distribution of seeds of valuable economic trees, and with the dissemination of information upon forestry matters.

The Microscopist makes investigations relating to parasitic growths ; to the characteristics of fibres, and to the adulteration of foods.

The Seed Division collects new and valuable seeds and plants for propagation in this country and distributes them to applicants, who are required to furnish the department with a report as to results obtained with seeds so furnished them.

The publications of the Department of Agriculture consists of an

Annual Report.

Special Reports on various subjects, published from time to time.

Bulletins by the Divisions of
Botany, Chemistry, Statistics, Entomology, Forestry, Pomology and Experiment Stations.

Periodical Bulletins entitled :
"Insect Life," "North American Fauna," "Journal of Mycology," and "Contributions from the U. S. National Herbarium."

The Weather Bureau.

Professor MARK W. HARRINGTON, Chief.

The Weather Bureau, which was transferred to the Department of Agriculture on July 1, 1891, has its office at the corner of Twenty-fourth and M streets northwest, immediately adjoining the grounds of the Columbia Hospital.

The Library, under the management of Mr. O. L. Fassig, containing 11,000 volumes and 3,000 pamphlets; the Instrument Room, under Professor C. F. Marvin, and the Indications Room will be found interesting to visitors.

The observations made daily at 8 a. m. are displayed on a printed map with accompanying predictions for the next thirty-six hours, and will be furnished by 11 a. m. daily for the use of the American Association, the Geological Society of America, and the International Congress of Geologists.

Post-Office Department.

Honorable JOHN WANAMAKER, Postmaster-General.

This department occupies a massive structure opposite the Department of the Interior. It covers an entire square bounded by E and F and Seventh and Eighth streets. It is built of white marble. The main feature of interest is the dead-letter office, to visit which a pass from the Chief Clerk is necessary.

Department of Justice.

Honorable WILLIAM H. H. MILLER, Attorney-General.

This department is situated on Pennsylvania avenue, between Fifteenth street and Lafayette Square. It is four stories high and built of Potomac Seneca redstone. The office of the Attorney-General contains a gallery of portraits of all the Attorneys-General of the United States since the foundation of the government. The Court of Claims occupies the first floor of the building.

The Smithsonian Institution.

Professor S. P. LANGLEY, Secretary.

The Smithsonian Institution is supported by a permanent fund at present amounting to $703,000, the accumulations of a bequest to the United States made in 1826 by James Smithson, a scientist of England, "to found at Washington under the name of the Smithsonian Institution an establishment for the increase and diffusion of knowledge among men." Some years were occupied in securing the bequest and in perfecting plans for carrying out its provisions. By Act of Congress. August 10, 1846, the Institution was created as an "Establishment," of which the President and the other principal officers of the general government were made ex-officio members, while the direction of affairs was intrusted to a Board of Regents "to be composed of the Vice-President, the Chief Justice of the Supreme Court, (the Mayor of Washington), three members of the Senate, and three members of the House of Representatives, together with six other persons other than members of Congress, two of whom shall be resident in the City of Washington, and the other four shall be inhabitants of some state, but no two of the same state."

The plan of organization adopted contains the following propositions :

"I. To increase knowledge. It is proposed to stimulate men of talent to make original researches by offering suitable rewards for memoirs containing new truths.

"II. To increase knowledge. It is also proposed to appropriate a portion of the income annually to special objects of research under the direction of suitable persons.

"III. To diffuse knowledge. It is proposed to publish a series of periodical reports giving an account of the progress of the different branches of knowledge.

"IV. To diffuse knowledge. It is proposed to publish occasionally separate treatises on subjects of general interest."

A further part of the plan contemplated the formation of a Library, a Museum and a Gallery of Art.

While the developments of the past forty-five years have been greater in some directions than in others, the original plan has been consistently followed with highly gratifying results.

The chief administrative officer of the Institution is the Secretary, a position which has been occupied by only three persons, namely, Joseph Henry, Spencer F. Baird and Samuel P. Langley. The Assistant Secretary is the officer in charge of the National Museum.

The Smithsonian Building is situated in that division of the Mall, between Seventh and Twelfth streets, known as Smithsonian Park. It was built, 1847-1856, at a cost of $450,000, after designs by Renwick. The style is termed "Norman" or "Romanesque," and the material is a lilac-gray freestone, found in the red sandstone formation about twenty-three miles above Washington. The building contains at present the administrative offices, reading room, the exchange department, and several collections of the National Museum, notably those of birds, shells and archæological specimens.

The Library of the Smithsonian Institution consists of more than 250,000 volumes and parts of volumes. It is for the most part deposited in the Congressional

Library, but each department of the Institution and the National Museum is supplied with such books as relate to its special work. The collection of the publications of scientific societies and of scientific periodicals is very large.

The Smithsonian Bureau of International Exchanges, which was early instituted, has accomplished a great work in distributing in this country and abroad the government publications, and the publications of scientific and literary societies of almost every country in the world. By its agency the Smithsonian Library has been enriched with many rare works of reference, and the publications of the Institution have been scattered far and wide. The general government has now assumed the support of this Bureau, and has made the Institution its agent in distributing all government scientific publications to foreign countries. An idea of the magnitude of the work may be formed from the statement that more than 90,000 packages, representing over 100 tons of books, pass through the Bureau each year. Over 16,000 correspondents, societies and individuals, are upon the exchange list.

The Smithsonian Institution is charged by Congress with the expenditure of the sums annually appropriated for the Bureau of International Exchanges, the Bureau of Ethnology, the National Museum, and the National Zoological Park.

PUBLICATIONS.—The Smithsonian Institution has three classes of publications :

First—"Contributions to Knowledge," a quarto series, in which are included memoirs giving new facts obtained in original research.

Second—"Miscellaneous Collections," an octavo series, containing practical papers or treatises, such as systematic lists of species in the animal, vegetable or mineral kingdoms, tables of natural constants, scientific bibliographies, and other summaries.

Third—"Annual Reports," an octavo series, containing the yearly report of the Secretary to Congress of work done, and supplemented by short papers upon the most important scientific discoveries of the year, by bibliographies of current literature, and by accounts of progress in various sciences.

In the Park near the northwestern corner of the building is a bronze statue to the memory of Joseph Henry, the first Secretary, to whose wise guidance the Institution owes a large share of its prosperity.

The National Museum.

G. BROWN GOODE, Assistant Secretary.

The National Museum is maintained by annual Congressional appropriations which are expended under the direction of the Smithsonian Institution, and the Assistant Secretary of the latter is in charge of the Museum. The Museum originated in 1840, when the National Institution was organized, and the collection of the Wilkes expedition constituted its nucleus. In 1849 a museum was established by the Smithsonian

Institution, and this, in 1858, was made the repository of all the scientific collections of the government, including those of the National Institution. It acquired very large collections from various sources at the close of the Centennial Exposition, in 1876, and from that time has been recognized as the National Museum of the United States. The large accessions in 1876 led to the erection of the present museum building (1879–1881), but the additions since its occupation are sufficient to fill a much larger building than the present one. Out of thirty-three departments and sections there are seven to which no room for exhibition purposes can be assigned in the Museum building for lack of space. To some of these departments, however, have been allotted inadequate accommodations in the Smithsonian building.

No official guide to the collections has yet been published, although the curators of several of the departments have prepared hand-books descriptive of the collections under their charge. On the right, at the entrance to the Museum, is a bureau of information for the guidance of visitors.

The following is a list of the Scientific Departments in the Museum :

I. †Arts and Industries ‖ : G. Brown Goode,* Assistant Secretary, acting as curator.
II. †Ethnology : Otis T. Mason, curator.
　† American Aboriginal Pottery : Wm. H. Holmes,* curator.
III. ‡ Prehistoric Anthropology : Thomas Wilson, curator.
IV. † Mammals : F. W. True, curator.
V. ‡ Birds : Robert Ridgway, curator.
　‡ Birds' Eggs : Capt. Charles E. Bendire,* curator.
VI. 　Reptiles : Dr. Leonhard Stejneger, curator.
VII. ‡ Fishes : Tarleton H. Bean,* curator.
VIII. † Vertebrate Fossils : O. C. Marsh,* curator.'
IX. ‡ Mollusks : W. H. Dall,* curator.　R. E. C. Stearns, adjunct curator.
X. † Insects : C. V. Riley,* curator.
XI. 　Marine Invertebrates : Richard Rathbun,* curator.
XII. † Comparative Anatomy : Frank Baker,* curator.
XIII. † Invertebrate Fossils :
　　　Palœozoic—C. D. Walcott,* curator.
　　　Mesozoic—C. A. White,* curator.
　　　Cenozoic—W. H. Dall,* curator.
XIV. † Fossil Plants : Lester F. Ward.* curator.
XV. § Botany : Dr. George Vasey,* curator, Botanist of the Department of Agriculture.
XVI. † Minerals : F. W. Clarke,* curator.
XVII. † Geology : George P. Merrill, curator.

‖ This department at the present time includes twelve different sections, each of which is under the charge of a curator, or an assistant acting as a curator.
* Honorary.
† Departments with exhibits in the Museum building.
‡ Departments with exhibits in the Smithsonian building.
§ The National Herbarium is for the present kept in the building of the Department of Agriculture.

For information regarding the general collections of the National Museum the visitor is referred to a guide : "The Smithsonian, the National Museum and the Zoo," to be purchased (25c.) in the rotunda of the Museum. This book is not an official publication. For the geological collections, the arrangement of which has recently been changed, the visitor should secure the preliminary hand-book of the department of geology by the curator, G. P. Merrill. The Geological Department embraces both economic and general geology. In the Mineralogical Hall are the systematic mineral collection, a collection of gems and precious stones and one of meteorites.

The publications of the National Museum embrace the "Proceedings," the "Bulletins" and the "Annual Report," which forms the second volume of the Smithsonian Report, and whose appendix contains many scientific papers.

Bureau of Ethnology.

Major J. W. POWELL, Director.

The Bureau of Ethnology was organized in 1879, and was placed under the direction of Major J. W. Powell, Director of the Geological Survey. In its early years it was so closely associated with the Geological Survey that its work was and still is often confounded with the work of that Bureau. It is, however, a separate and distinct organization supported by specific appropriations made by the general government, and the general supervision of its scientific work is confided to the Secretary of the Smithsonian Institution. The appropriation for the current year is $50,000.

The work of the Bureau comprises the whole field of North American Ethnology, including Archæology ; and the range of its work extends from Alaska on the north to Panama and the Isthmus of Darien on the south. Its collections are deposited in the National Museum, and those branches of Indian art to which it has especially devoted attention are now illustrated by collections of specimens which compare favorably with those of the largest museums. Its collection of aboriginal American Pottery, now in the National Museum, is notably the largest and finest in existence.

The publications of the Bureau comprise Annual Reports, to which are appended papers upon subjects of general interest, a series of Bulletins, consisting of reports upon special subjects, and Quarto Contributions to North American Ethnology. These publications are distributed through the exchange system of the Smithsonian Institution.

The office of Major Powell is in the Geological Survey Building, No. 1330 F street northwest.

The U. S. Commission of Fish and Fisheries.

Colonel MARSHALL McDONALD, Commissioner.

The Commission was established primarily with the object of determining the cause of decrease among food-fishes, and of suggesting measures for the improvement of the fishing grounds. Its scope, however, has been rapidly enlarged to cover all matters pertaining to fisheries which come within the jurisdiction of the general government, including the propagation of useful fishes and the methods and statistics of the fishing business. Colonel Marshall McDonald, the present Commissioner, succeeded Professor Spencer F. Baird, upon the death of the latter in 1887.

The work of the Commission is arranged under three divisions, as follows : The Division of Scientific Inquiry is charged with the investigation of the fishing grounds relative to their resources and characteristics, their depletion and the methods suited to their replenishment ; and also with the study of the habits and development of fishes as a basis for fish culture, legislation and fishery methods. The Division of Fish Culture undertakes the propagation of food-fishes, their distribution to different localities, the restocking of exhausted grounds, and the introduction of useful foreign species. The Division of Fisheries considers the methods and apparatus of the fishermen with a view to their improvement, and collects the statistics of the different branches of the business.

The investigations along the seacoasts are chiefly carried on by means of two steamers, the Albatross and Fish Hawk, and one sailing vessel, the schooner Grampus. The Albatross is now stationed on the Pacific coast, the Fish Hawk and Grampus on the Atlantic coast ; the two latter vessels being also employed to some extent in fish culture. There are two marine stations for the hatching of cod, mackerel, lobsters and several other salt-water species, one located at Wood's Holl, the other at Gloucester, Massachusetts. The former is also adapted to scientific inquiries, being provided with large and well equipped laboratories for biological and physical research. A number of fresh-water and anadromous fishes are propagated upon a very exhaustive scale, the most important being the shad, lake whitefish, carp, Atlantic and Pacific salmon and several species of trout. For conducting this work twenty-one stations have been established in different parts of the country, each embodying the most approved methods applicable to the branch of fish culture for which it is adapted. Several cars, specially constructed for that purpose, are used for the distribution of the eggs and fry as well as the adult fishes.

The officers of the Commission are located in Armory Square, Washington (B street southwest, between Sixth and Seventh streets). The same building contains a biological laboratory, extensive aquaria for the study and display of salt and fresh-water fishes, and also one of the principal shad-hatching stations, for which the supply of eggs is obtained from the important fisheries of the Potomac river during the spring.

Large ponds for the breeding of German carp are situated on the Mall near the Washington Monument. Tench, golden ide and goldfish are also produced there in small numbers, and one of the ponds now contains about 2,000,000 shad fry of the last season's hatching.

Collections illustrating the work of the Fish Commission are exhibited by the National Museum. The models of fishing boats, fishing apparatus and the Cetaceans are displayed in the Museum building, and the fishes, mollusks, crustaceans and lower marine invertebrates in the Smithsonian building, where a large part of the zoological material obtained during the investigations of the Commission is also stored.

Educational Institutions.

Georgetown University is the oldest educational institution of the Catholic Church in America. Founded in 1789 ; incorporated as a university in 1815. Has collegiate, law and medical departments. President, Rev. J. Haven Richards, S. J.

The Columbian University was incorporated by Act of Congress February 9, 1821, as a college and re-incorporated as a university in 1873. It has collegiate, law and medical departments. Its main building is that within which the meetings of the Congress of Geologists are held, corner of H and Fifteenth streets northwest. President, Dr. J. C. Welling.

Howard University is devoted to the higher education of the colored race. It was founded in 1867, and is supported by the Government. It has a collegiate department, and schools of theology, law and medicine. The average attendance is 300. President, Rev. J. E. Rankin.

Catholic University of America. Founded in 1889. Situated at Brookland, a suburb of the city, east of the Soldiers' Home. Is reached by the Metropolitan Branch of the Baltimore and Ohio Railroad. The Divinity School is the only department at present organized. The Rector is the Rt. Rev. John J. Keane, D.D.

Columbia Institution for the Deaf and Dumb and National Deaf Mute College. This institution has two departments, a primary and a collegiate ; the former established in 1857, the latter in 1864. It is supported by Congressional appropriations. The development of the institution has been from the first under the guidance of Dr. E. M. Gallaudet, now President of the Faculty. This college is the only one in the world for deaf mutes. It is situated just beyond the northeastern boundary of the city in the park called Kendall Green, a portion of the estate of Amos Kendall, the original promoter of the school and its first President.

Government Printing Office.

THIS building is situated on the corner of North Capitol and H streets. It is 300 feet long on H street and four stories high. All the printing and binding ordered by the Legislative, Executive and Judicial Departments of the Government is done in this building. It is the largest establishment of the kind in the world. Open from 8 a. m. to 5 p. m.

—

Libraries of Washington.

THE Libraries of the General Government have arisen from the exigencies of public business, and with the growth of new bureaus the formation of separate reference libraries has become necessary. With few exceptions these libraries have been formed with reference to the special need of bureaus, and though small are very complete in their own subjects.

Library of Congress.

THE Library of Congress dates from the first meeting of Congress in the City of Washington in 1800 ; it was burned by the British in 1814 ; was replaced by the purchase of Jefferson's Library and grew to contain about 55,000 books in 1851, when a fire destroyed all but 20,000 books. Since 1852 it has grown steadily and of late rapidly. In 1866 the books accruing to the Smithsonian Institution by exchange were diverted to the Library of Congress, and in 1867 the large historical collection of Peter Force was purchased and added to it. It now numbers about 650,000 volumes.

House of Representatives.

THE Library of the House of Representatives is almost exclusively of a documentary character, containing legislative and executive volumes for the use of members of the House. Including duplicates it numbers 125,000 volumes.

Senate.

THE Library of the Senate was begun in 1852, and consists entirely of public documents for the use of Senators. At present it contains 47,000 volumes.

Executive Mansion.

THE Library of the Executive Mansion is very like a miscellaneous family library. It began to accumulate in the time of President Madison and now contains about 4,000 volumes.

State Department.

THE Library of the State Department dates from the organization of the government, in 1789. It is made up of works on the laws of nations, diplomatic and general history, voyages and cognate subjects, and contains 50,000 volumes and 3,000 pamphlets.

The BUREAU OF AMERICAN REPUBLICS has collected about 1,100 volumes relating to the Spanish republics of this continent, with special reference to all questions of international comity and commerce.

Treasury Department.

THE General Library of the Treasury is for the entertainment of Treasury Department clerks and is mainly biography, history and fiction. It contains 18,000 volumes.

The BUREAU OF STATISTICS began in 1866 to collect the statistical publications of the world, and now contains 5,000 volumes and 6,500 pamphlets.

The COAST SURVEY Library contains about 8,000 volumes and 7,000 pamphlets of highly special character. Its archives contain about 5,000 original manuscript maps and 65,000 record books of observation, computation and reduction. Its collection of foreign maps and charts numbers 9,000.

The LIGHT-HOUSE BOARD has a library begun in 1852, and containing now 3,496 volumes on light, sound, naval architecture and engineering.

The MARINE HOSPITAL BUREAU has a library of 1,500 books and 1,000 pamphlets.

War Department.

THE General Library of the War Department was begun in 1832 under Secretary Lewis Cass. It is devoted chiefly to military science and contains 30,000 volumes.

The Library of the ORDNANCE BUREAU is devoted to military engineering, gunnery and military and civil law. It contains 3,000 volumes.

The Library of the SURGEON-GENERAL's OFFICE has been formed since the war of 1861-'65, and is practically the medical section of the Library of Congress. It covers the entire field of medical and surgical literature, and contains 101,969 volumes and 152,225 pamphlets.

The SOLDIER's HOME has a library dating from 1850. It is of a miscellaneous character and contains 5,632 volumes.

The Library at the ARMY HEADQUARTERS, begun by General Grant and added to by Generals Sherman and Sheridan, is of considerable value for its especial purpose.

Navy Department.

THE General Library of the Navy Department is made up of historical, scientific and legal works with especial relation to naval affairs. It numbers 24,086 volumes and 1,000 pamphlets.

The BUREAU OF MEDICINE AND SURGERY has a library of special reference works of a medical and scientific character, which numbers 15,998 volumes.

The HYDROGRAPHIC OFFICE library was begun in 1867, and is made up of hydrographical, nautical and meteorological works ; it contains about 3,000 volumes and 2,000 pamphlets.

The Library of the NAVAL OBSERVATORY dates from the founding of the Observatory in 1843. It is a collection of the best works relating to astronomy, mathematics and geodesy, and numbers 13,000 volumes and 3,000 pamphlets.

Post=Office Department.

BUT a small portion of the library of the Post-Office Department is general literature. It consists of public documents pertaining to the duties of the office, and numbers 8,000 volumes.

Interior Department.

THE Library of the Interior Department was begun in 1850, and is made up of miscellaneous literature for the use of Department clerks. It has 10,500 volumes.

The Library of the BUREAU OF EDUCATION was begun in 1870, and contains books and journals on educational topics, and school reports of all the world, to the number of 17,500 volumes.

The GENERAL LAND OFFICE Library contains the laws and documents relating to the public domain, and numbers 3,000 volumes.

The COLUMBIA INSTITUTION FOR THE DEAF AND DUMB has a collection of works relating to the instruction of the deaf and dumb surpassed by only one other in the world. It numbers 4,000 volumes.

The Scientific Library of the PATENT OFFICE was begun in 1839, and contains a very fine collection of works in all departments of science and all reports needed for reference in determining questions concerning inventions. It numbers 50,000 volumes.

The Library of the GEOLOGICAL SURVEY is not yet ten years old, but has already a practically complete collection of official geological reports and of the standard works on geology and its cognate subjects to the number of 30,000 volumes, 40,000 pamphlets and 22,000 maps.

Department of Justice.

THE Library of the Department of Justice was begun in 1853, and forms an excellent collection of American, English, Spanish-American and Roman law books. It contains 20,000 volumes.

The Library of the SOLICITOR OF THE TREASURY dates from 1843, and is made up wholly of law books and official documents for reference to the number of 7,000 volumes.

Department of Agriculture.

THE Department of Agriculture has a collection of works on agriculture and natural history, and their kindred branches, to the number of 24,000 volumes and 8,000 pamphlets.

The Library of the WEATHER BUREAU was begun in 1871. It is made up entirely of books on meteorology, telegraphy and cognate subjects to the number of 12,000 books and 2,500 pamphlets.

These libraries of the General Government contain more than 1,248,761 books and 228,225 pamphlets, most of which are accessible to any student in legitimate scientific study.

Society Libraries.

Among important libraries not governmental should be noticed the following : The American Medical Association Library, which contains 7,000 volumes, the Law Library of the Bar Association, which numbers 7,000 books, the Library of the Supreme Council 33°. a collection especially rich in works of history, religion, philosophy and folk-lore to the number of 15,000, which though especially intended for and free to all masons is yet accessible to every student : the Masonic Library of 3,000 volumes and the library of the Young Men's Christian Association numbering 2,000 books.

School Libraries.

CARROLL INSTITUTE has a select library of 3,000 volumes : Columbian University has a miscellaneous collection of 6,000 books and 2,000 pamphlets ; Georgetown College possesses the fine Riggs Library of 35,000 volumes and of very broad scope ; Gonzaga College and St. John's College have special libraries of 10,000 and 4,000 volumes respectively ; and Howard University has 15,000 books, among which are some rare Americana.

A general table of Washington libraries is here given :

30

Washington Libraries.

	Books.	Pamphlets.
Academy of the Visitation, - - - - -	1,000	
American Medical Association. - - -	7,000	
Bar Association, - - - -	7,000	
Bureau of Education—Gov't, - - - -	17,500	
Bureau of Medicine and Surgery—Gov't, - -	15,998	
Bureau of Ordnance (Navy Dept.) Gov't,	3,000	
Bureau of Statistics (Treas. Dept.)—Gov't, - -	5,000	6,500
Carroll Institute, - - - - - -	3,000	
Coast and Geodetic Survey—Gov't, - - -	8,000	7,000
Columbia Institution for Deaf and Dumb,	4,000	
Columbian University, - - - - -	6,000	2,000
Department of Agriculture—Gov't, - -	24,000	8,000
Department of Justice—Gov't, - - -	20,000	
Department of State—Gov't, - - - -	50,000	3,000
Department of the Interior—Gov't, - - -	10,500	
District of Columbia—Gov't, - - -	2,000	
Executive Mansion—Gov't, - - - -	4,000	
General Land Office—Gov't, - -	3,000	
Geological Survey—Gov't, - - -	30,000	42,000
Georgetown College, (Riggs Library), - -	35,000	
Gonzaga College, - - - - - -	10,000	
Government Hospital for the Insane—Gov't, -	2,480	
Health Department, D. C.—Gov't, - -	2,000	
House of Representatives—Gov't, - - -	125,000	
Howard University, - - - - -	15,000	
Hydrographic Office—Gov't, - - - -	3,000	2,000
Library of Congress—Gov't, - - - -	650,000	200,000
Library of Supreme Council 33° southern jurisdiction U. S. A., - - - - - -	15,000	
Light Battery C, 3d Artillery - - - -	2,000	
Light-House Board (Treas. Dept.)—Gov't, -	3,496	
Marine Hospital Bureau—Gov't - - -	1,500	1,000
Masonic Library, - - - - - -	3,000	
Nautical Almanac Office—Gov't, - - -	1,600	
Naval Observatory—Gov't, - - - -	13,000	3,000
Navy Department—Gov't, - - - -	24,086	1,000
Patent Office Scientific Library—Gov't, - -	50,000	
Post-Office Department—Gov't, - - -	8,000	

31

Washington Libraries (Continued.)

	Books.	Pamphlets.
St. John's College, - - - - -	4,000	
Senate—Gov't, - - - - -	47,000	
Soldiers' Home—Gov't, - - - -	5,632	
Solicitor of the Treasury—Gov't, - - -	7,000	
Surgeon General's Office, U. S. Army—Gov't,	101,969	152,225
Treasury Department—Gov't, - - - -	18,000	
War Department—Gov't, - - - - -	30,000	
Weather Bureau—Gov't, - - - -	12,000	2,500
Young Men's Christian Association. - - -	2,000	
Total, - - - - - - -	1,362,761	230,225

Private Libraries.

THE existence of this vast body of literature in the city has naturally operated against the formation of great private libraries in Washington, but there are nevertheless some worthy of notice.

The historical library of the late George Bancroft, the general libraries of Justice Joseph Bradley, Justice Horace Gray, Mr. Henry Adams, Col. John Hay and Mr. John G. Nicolay, the musical library of Mr. Edward Clarke, the Scotch library of Mr. Wm. R. Smith, the library of Americana of Mr. L. A. Brandenburg, and the collection of books relating to the civil war of 1861–'65 which Mr. John Davenport has collected, are very fine in their class.

One of the interesting collections in the city is the one made by Mr. Frederick Schneider who, in the intervals of a life as an iron founder and dealer in hardware, has through correspondence with booksellers of Europe collected a library of illustrated books, from the Nuremburg Chronicle to the present day, which contains rarities not in the great libraries. He has printed an annotated catalogue of his treasures, setting the type and doing all the press-work, etc., with his own hand.

General Information.

A BUREAU OF INFORMATION will be maintained during the sessions of the Congress in the Columbian University, where some one will be in constant attendance. Programs, circulars, the Washington Directory, railroad guides and timetables, etc., may be found here. Macfarlane's Geological Railway Guide and local guide-books to Washington may be purchased at the bureau.

The *Telephone* is free for use of members by courtesy of S. M. Bryan, President of the Chesapeake and Potomac Telephone Company. The District Messenger call may be used for messengers, cabs, etc., and telegraph.

There will be a temporary *Post-Office* in the building, where mail for members will be found. Stamps can be purchased here, and the Postal Guide consulted.

Money Exchange.—Foreign members of the Congress desiring to exchange foreign currency can do so at the banking house of Crane, Parris & Co., No. 1344 F street northwest. Arrangements will also be completed whereby this exchange can be made at the office of the Congress in the Columbian University.

Scientific Societies of Washington.

Philosophical Society,	Organized 1871.
President : T. C. Mendenhall.	
Anthropological Society,	Organized 1879.
President : J. C. Welling.	
Biological Society,	Organized 1880.
President : C. Hart Merriam.	
Chemical Society,	Organized 1884.
President : R. B. Warder.	
Microscopical Society,	Organized 1884.
President : Thomas Taylor.	
Entomological Society,	Organized 1884.
President : George Marx.	
National Geographic Society,	Organized 1888.
President : Gardiner G. Hubbard.	
Women's Anthropological Society,	Organized 1885.
President : Alice C. Fletcher.	

Offices of Foreign Legations.*

† *Austria-Hungary :* 1537 I street northwest.
Chevalier de Tavera, Envoy Extraordinary and Minister Plenipotentiary.
† *Belgium :* 1336 I street northwest.
Mr. Alfred Le Ghait, Envoy Extraordinary and Minister Plenipotentiary.
France : 1901 F street northwest, (two squares west of the State, War and Navy
Building).
Mr. Theodore Roustan, Envoy Extraordinary and Minister Plenipotentiary.
† *Germany :* 734 Fifteenth street northwest, (opposite the Columbian University).
Count Ludwig von Arco-Valley, Envoy Extraordinary and Minister Plenipotentiary.
§ *Great Britain :* Corner Connecticut avenue and N street northwest.
Sir Julian Pauncefote, Envoy Extraordinary and Minister Plenipotentiary.
† *Italy :* 729 Eighteenth street northwest.
Marquis Imperiali di Francavilla, Chargé d'Affaires.
† *Mexico :* 1413 I street northwest.
Senor Don Matias Romero, Envoy Extraordinary and Minister Plenipotentiary.
Netherlands : Office of the Consulate-General of the Netherlands, New York City.
Mr. G. de Weckherlin, Envoy Extraordinary and Minister Plenipotentiary.
§ *Russia :* 1705 K street northwest.
Mr. Alexandre Greger, Chargé d'Affaires.
Spain : 1400 Massachusetts avenue northwest.
Senor Don Miguel Suarez Guanes, Envoy Extraordinary and Minister Pleni-
potentiary.
§ *Sweden and Norway :* 2011 Q street northwest.
Mr. J. A. W. Grip, Envoy Extraordinary and Minister Plenipotentiary.
§ *Switzerland :* 2014 Hillyer Place northwest.
Mr. Alfred de Claparéde, Envoy Extraordinary and Minister Plenipotentiary.

Clubs.

Cosmos Club—H street, opposite The Arlington.
Metropolitan Club—Corner of Seventeenth and H streets northwest.
University Club—No. 1726 I street northwest.
United Service Club—G street, near Seventeenth street northwest.

---

* Only those countries are given from which members are in attendance at the Congress.
Those marked † are within three squares of the Columbian University.
Those marked § are easily reached by the cars passing the Columbian University on H street.

Hotel Accommodations.

Special rates have been secured for members of the Congress at the following hotels, which are within five minutes' walk of the Columbian University.

Arlington Hotel (Headquarters)—On Vermont avenue from H to I streets. (American plan). Regular rate $5 per day and upwards, according to accommodations. A reduction of one-third of these rates will be allowed to members of the Congress.

The Arno—On Sixteenth street, between J and K streets. (European plan). Rooms at $1 each for members of the Congress, including use of the hotel baths. Private bath-rooms $1 per day extra. Restaurant and café in the hotel.

Ebbitt House—Corner of F and Fourteenth streets. (American plan). Adjoining the offices of the U. S. Geological Survey, Regular rates $4 per day. Rates to members of the Congress $2.50 per day and $1 extra for rooms with bath.

The Elsmere 1408 H street, between Fourteenth and Fifteenth streets. Board and lodging for members of the Congress at $10.50 per week during the meetings.

Arrangements have been made to provide apartments in lodging-houses for such as may desire them.

Drives Around Washington.

The Soldiers' Home.—This is one of the most attractive drives in the suburbs of the city. The grounds are beautifully laid out and are kept up as a park. President Lincoln resided here in the summer during his administration. It is three miles from the Arlington Hotel.

Arlington and Fort Myer, situated on Arlington Heights, overlooking the Potomac. The former was the home of George Washington Custis, and in later years was the residence of General Robert E. Lee. The estate was sold under the confiscation act of 1863, and 200 acres set apart as a National Cemetery. Over 16,000 soldiers lie buried here. General Sheridan's grave is but a short distance from the house. The drive is through Georgetown and over the Aqueduct Bridge. From the portico the view of the Potomac Valley is exceptionally fine and adds much to the pleasure of the drive. Distance from Washington, five miles.

An attractive drive is through the northwest portions of the city to the *Zoological Park*, thence northward to the country in and adjacent to the new *Rock Creek Park*.

Still another drive is to follow the Conduit Road along the north and east side of the Potomac River to *Glen Echo Heights* and *Cabin John Bridge*. The bridge is a magnificent structure spanning Cabin John Run ; it is 20 feet wide, with an extreme length of 420 feet. It is said to be the largest single span stone arch in the world. At the hotel near the bridge one can obtain a good dinner.

R. L. Cooper, No. 1335 H street northwest, offers the following special reduced rates for carriages to members of the Congress :

To Soldiers' Home and return, $3.

To Arlington and Fort Myer and return, $4.

To Cabin John Bridge and return, $5.

William F. Downey, No. 1624 L street northwest, and B. F. McCaully & Co., No. 920 O street northwest, also offer reduced rates for carriages.

35

Excursions in the Neighborhood of Washington.

Mount Vernon—the former residence and now the Tomb of Washington—situated on the Potomac river, ten miles below the Capitol, is easily reached by steamer which leaves daily, except Sunday, at 10 a. m. The boat reaches the city on its return trip at 2.20 p. m. Fare, round trip, $1, including admission to Mount Vernon.

Old Point, Fort Monroe, and Virginia Beach. These points, on the Virginia shore, are reached by the new steamers of the Washington and Norfolk line, which leave Washington daily at 7 p. m., passing the evening and night on the Potomac river and Chesapeake bay, arriving at Old Point at 7 a. m. the following day. Here the time may be pleasantly spent in visiting the Hygeia Hotel, Fort Monroe, and the Soldiers' Home at Hampton. Those remaining on the steamer reach Norfolk at 8 a. m., where the day may be passed in visiting the city, the U. S. Navy Yard at Portsmouth, or taking a short trip by rail to Virginia Beach, on the Atlantic. Returning steamers leave Norfolk at 5 p. m., Old Point at 6 p. m., arriving in Washington at 7 a. m. the following day. Fare, round trip, $5. Staterooms, $1 and $2 each way, according to location.

Luray Cavern, Virginia, situated about one mile west of Luray station, on the Shenandoah Valley Railroad, and sixty-five miles from Washington, is reached by the Baltimore and Ohio Railroad, connecting at Shenandoah Junction for Luray. The best excursion is that leaving Washington from the Baltimore and Ohio Depot at 3.30 p. m., arriving at Luray at 7.45 p. m. Visit the Cavern that evening, after supper at Luray Inn. Leave Luray the following day at 7.10 a. m., arriving in Washington about 11.45 a. m. Those wishing to see Harper's Ferry and vicinity can stop over and find trains to Washington at 3.06 p. m., 4.25 p. m. and 6.18 p. m. Fare, Washington to Luray and return, $5.50. Admission to the Cavern, $1.00. Board at Luray Inn, $2.00 and $2.50 per day.

Excursions after the Congress.

Members of the Congress will have received the Congress Circular giving the itinerary of the long excursion which it is proposed to make from Washington to the Yellowstone Park, Salt Lake, Denver, and back via Chicago and Niagara Falls to New York, starting September 2d, and to be en route twenty-five days. The expense of this trip will be $265.00 for each person.

Another excursion contemplated by the Congress Committee will leave Washington September 2d, and make a circuit through Pennsylvania, via Philadelphia, Pottsville, Wilkesbarre, Harrisburg and Cresson, visiting the Anthracite basins, and localities made famous by Rogers and Lesley as illustrating Appalachian stratigraphy, structure and topography. Glacial phenomena will be seen at Berwick. The production and use of oil and gas will be shown at Pittsburg. From Pittsburg the party will pass through the Connellsville coke region, the Valley of Virginia, stopping at Luray Cave; thence down the New river gorge to Pocahontas, and to Middleboro' at Cumberland Gap; return via Knoxville, across the Palæozoic of Tennessee and the Archæan of North Carolina, and up the coastal plain to Washington. The carrying out of this plan will depend upon the number wishing to make the trip. Railroads have cordially offered reduced rates, and mining companies opportunities for seeing the things of interest. The cost will probably fall below $100, and sixteen days will be required. A special descriptive circular and itinerary will be issued.

36

Street Car Lines.

Street car fare, 5 cents, or 6 tickets for 25 cents. Tickets of one line received for fare on all other lines

Transfer tickets can be obtained at points of intersection of lines belonging to the same Company. (See map.)

The green cars of the Metropolitan Railroad Company passing the Columbian University on H street go west to Georgetown, passing the Legation of Great Britain, and near those of Russia, Sweden and Norway, and Switzerland (see p. 34). They go east past the U. S. Geological Survey, Patent Office, Post-Office Department, Pension Building, Baltimore and Ohio Railroad depot, the Capitol, and out East Capitol street one mile to Lincoln Park. On Ninth street is another line belonging to this Company. These cars go north beyond the city limits and south past the Baltimore and Potomac Railroad depot to the wharves at the foot of Seventh street.

One block south from the University the cars of the Washington and Georgetown Railroad Company go west past the White House to Georgetown, and south past the Treasury Department to Pennsylvania avenue, thence east to the Capitol and Navy Yard. They pass near the depot of the Baltimore and Potomac (Pennsylvania) Railroad on Sixth street. By transfer to the cable cars on Seventh street one can go north to the city limits or south past the Fish Commission and Army Medical Museum to the steamboat wharves and the Arsenal grounds.

On Fourteenth street, one block east of the Columbian University, is another line of cars of the Washington and Georgetown Railroad Company. These go north to the city limits, while southward they join the main line at the Treasury Department, so that from this point to the Capitol the cars of the two lines alternate. The Fourteenth street cars leave Pennsylvania avenue at the foot of Capitol Hill and go to the Baltimore and Ohio depot.

RATES OF FARE FOR HACKS, CABS AND OTHER VEHICLES.

(Extract from Police Regulations).

BY THE HOUR.	Between 5 a. m. and 12.30 a. m.	Bet'n 12.30 a. m. and 5 a. m.
For one passenger or two passengers, for the first hour................	$0 75	$1 00
For each additional quarter of an hour or part thereof	20	25
Provided, That for multiples of one hour the charge shall be at the rate per hour of	75	1 00
For three or four passenger, for the first hour..........................	1 00	1 25
For each additional quarter of an hour or part thereof	25	35
Provided, That for multiples of one hour the charge shall be at the rate per hour of	1 00	1 25
BY THE TRIP.		
By the trip of fifteen squares or less for each passenger..........	25	40
For each additional five squares or part thereof	10	15
Provided, That for multiples of fifteen squares the charge shall be at the rate for each fifteen squares of	25	40

Two-horse hacks, for four persons, may charge $1.50 for the first hour, and 25 cents additional for each extra quarter hour.

RAILROADS.

Baltimore and Potomac (Pennsylvania) Railroad, } Depot : Corner Sixth and B streets northwest.
Richmond and Danville Railroad,
Chesapeake and Ohio Railroad,

Baltimore and Ohio Railroad—Depot : Corner New Jersey avenue and C street northwest.

The Geology of Washington and Vicinity.*

THE GENERAL PHYSIOGRAPHY.

There are in eastern United States three distinct physiographic provinces. Most conspicuous of these is the Appalachian zone, an area of long, low mountain chains of wonderful parallelism. At the eastern base of the mountains lies the Piedmont plateau, an undulating plain standing 500 to 1000 feet above sea level. Between this plateau and the ocean lies the Coastal Plain, a generally smooth lowland rising gently from ocean waters to altitudes reaching about 300 feet.

The rocks of the Appalachian zone are Paleozoic, running from the Carboniferous down to the Cambrian and probably to the Algonkian, aggregating 25,000 to 40,000 feet in thickness. The entire series is nearly or quite conformable ; the

* Prepared by W J McGee, with the collaboration of Professor G. H. Williams, and Messrs. N. H. Darton and Bailey Willis.

materials range from coal seams toward the summit, and pure limestone at various horizons, to coarse sandstones, and in Pennsylvania to great beds of conglomerate. The strata, originally horizontal or slightly inclined westward, have been deformed and altered in a variety of ways. In the western and central portions of the province they have been flexed symmetrically and thrown into a series of anticlinal and synclinal corrugations, seldom more than a mile or two in width though often scores or even hundreds of miles in length—a series of mountain-folds unparalleled elsewhere on the globe in length, symmetry, and concordance in direction. In the central part of the zone the symmetric flexing is combined with faulting, and in many cases the faulting is of that overthrust type which characterizes the Scottish Highlands and the Canadian Rocky Mountains. In the eastern portion of the zone the symmetric flexing fails, faulting (both normal and overthrust) prevails, and the rocks are more or less profoundly metamorphosed—the limestones transformed into marbles, the shales into slates, the sands into quartzites. Throughout the province the distinctive structure and the rock composition are both reflected in topographic configuration ; the prevailing forms are long narrow ridges, separated by long and generally narrow valleys ; but these land forms represent respectively the outcropping edges of hard strata and soft beds rather than original flexures.

The rocks of the Piedmont belt are more or less crystalline, chiefly metamorphic schists and gneisses of considerable diversity in composition, but sometimes including ancient eruptives, as well as quartz veins and dikes. The structure of the province is obscure and diverse, and has not yet been fully investigated. It is known, however, that in the latitude of Washington at least the Piedmont belt is separable into two distinct parts. Of these the western is composed of semi-crystalline slates, phyllites and schists having a constant inclination toward the east ; while the eastern part is made up, except for a few included folds of the less crystalline rocks, of highly crystalline gneisses and a variety of foliated eruptives, all of which have a prevailing dip toward the west. The nearly vertical position of strata intermediate between these extremes gives a pseudo fan-structure to a section of the Piedmont plateau in Maryland. The line between the western semi-crystalline and the eastern gneissic areas is not a sharp one ; and there is an apparent progressive increase in the intensity of metamorphism from the western border to the eastern limit of the Piedmont belt by which casual students have been misled. The surface of the zone is characterized by meandering stream channels and wandering divides, with moderately strong local relief ; yet, while the harder rocks of the province find a certain expression in the topography, the general configuration is independent of rock structure but represents baselevel conditions during past eons.

The composition and configuration of the Piedmont zone are locally diversified by considerable areas of Mesozoic rocks, commonly referred to the Triassic. These rocks are red sandstones and red or purple shales, with occasional beds of conglomerate. They are characterized by strong dips toward the Appalachian zone ; and they are frequently cut and sometimes interbedded with or overlain by contemporaneous or younger dikes and sheets of trap. In the northern part of the Coastal plain the trap occurs in considerable volume, and forms prominent ridges by which the topography of the entire Piedmont belt is dominated ; but in general the sandstones and shales are soft and friable, and find topographic expression in low-lying plains and basins.

The rocks of the Coastal plain are clastic, ranging in age from Pleistocene to middle Mesozoic, probably reaching a total thickness of 2,500 to 3,500 feet. The entire series inclines gently seaward, the inclination increasing from the newer to the older formations. The strata are manifestly made up of the debris of the Appalachian and Piedmont provinces, are rarely lithified, and range from alluvium or alluvium-like silts along the rivers and toward the coast, and glauconitic marls and fine clays in the middle of the series, to coarse gravels and beds of arkose toward the base and near the old shorelines. Except for the gentle inclination of the strata, and except for a dislocation coinciding with the inland margin of the province, the strata are not visibly deformed, but retain substantially the attitudes as well as the composition of original deposition. The surface of the province is commonly characterized by meandering rivers, throughout the middle Atlantic slope by broad estuaries, and in general by broad low divides, often terraciform—the configuration seldom expressing structure or localized earth movement, but representing simple erosion combined with wave action during several continental oscillations of general character.

The western boundary of the Appalachian zone is indefinite ; the characteristic corrugations gradually die out and the flexed strata of the Appalachian pass into the undisturbed strata of the interior plain.

The common boundary of the Appalachians and the Piedmont zone is generally trenchant, consisting of a prominent ridge of quartzite—the Blue ridge. Somewhat south of the latitude of Washington the ridge is simple and single ; where cut by the Potomac river west of Washington it is triple or quadruple ; in Maryland and Pennsylvania it is frequently multiple ; and in Virginia and the Carolinas it is sometimes interrupted and again divided ; but in general it definitely marks a fairly decided transition from comparatively simple to comparatively complex structure, and from incipient metamorphism to pronounced alteration in the rocks.

Throughout the middle Atlantic slope the common boundary of the Piedmont zone and the Coastal plain is pronounced ; along this line there is a sudden and decided transition in the rocks from highly altered crystallines to practically unaltered clastics ; along this line the water-ways change from narrow, rock-bound gorges of considerable declivity to broad tidal canals, and each river passes from the one province to the other in a cascade or rapid ; along this line the rivers are diverted from courses cutting across the trend of structure and athwart the provinces to courses parallel with the line of cascades, thus peninsulating most of the Coastal plain ; and along the line thus accentuated by the diverted drainage there is commonly a prominent scarp of Piedmont rocks overlooking the flat-lying rocks of the Coastal plain. This physiographic boundary is one of the most trenchant on the surface of the globe, and the natural line is emphasized by a prominent cultural line to which it gave origin ; all the principal cities of the eastern United States from New York to the Carolinas are located along this natural landmark.

The eastern boundary of the Coastal plain may be drawn at the shore of the Atlantic ; but it may more properly be drawn 100 miles off shore at the great submarine escarpment, 3,000 to 10,000 feet high, hugged by the Gulf Stream— in general configuration, in inclination of the surface, and unquestionably in structure and composition, the subaerial and the submarine portions of the Coastal plain are essentially a unit, and the present coast line is but an accident of present relation between sea and land.

Despite the diversity in rocks, structure and configuration in the three provinces, the principal rivers of the middle Atlantic slope traverse all alike. The Mohawk and the Hudson run around the northeastern extremity of the typical Appalachian zone, separating the three distinctive provinces from the analogous (but probably not homologous) physiographic tract of New England ; the Delaware, with its great secondarys, the Lehigh, the Susquehanna, the Potomac and the James, rise well within the Appalachian zone, cut through the successive ridges in a series of clefts, cross directly the Piedmont plateau, and, although diverted at the fall line, thence intersect the Coastal plain to the Atlantic ; and except at the fall line their courses are essentially independent of structural conditions. Yet even along the great rivers the boundaries of the physiographic divisions find expression : The Appalachian-Piedmont boundary is marked by narrow notches in the Blue Ridge, forming the far-famed " water gaps " of the Delaware, of the Lehigh, of the Susquehanna near Harrisburg, of the Potomac at Harper's Ferry, and of the James at Balcony Falls ; the Piedmont-Coastal boundary is still more strongly marked by the line of cascades on every river, large and small, from the Raritan in New Jersey to the Roanoke in North Carolina, and by the deflection of the water-ways which peninsulate the lowland plain from New York to Richmond.

THE LOCAL PHYSIOGRAPHY.

The City of Washington, like the other metropoles of the middle Atlantic slope, is located at the common boundary of the Piedmont and Coastal zones. The western part of the city is built on the ancient crystallines, the eastern on the non-lithified clastics ; though outliers of the clastic formations occasionally occur on the uplands some miles farther westward. Located like neighboring metropoles at the head of navigation, the city marks the position of the fall line. At Washington the Potomac river is tidal, and perhaps half a mile wide ; within four miles up stream the channel contracts at ordinary stages to barely 100 feet, changing meantime from a slack-water canal into a rushing torrent. This is the "Little Falls of the Potomac." Then follow twelve miles of nearly continuous rapids to the "Great Falls of the Potomac," where at ordinary stages the river contracts to about 50 feet and descends 40 feet in a succession of plunges of which the highest is about 15 feet. Between Great Falls and Washington the river occupies a narrow gorge excavated in a broader one, whose bottom averages 150 feet above tide; above Great Falls the river wanders over the bottom of the older gorge.

Just west of the city the embouchure of the gorge expands, and its walls merge into the general Piedmont scarp overlooking the Coastal lowland. Just east of the city lies Anacostia river, a goodly mill-stream only, clear and rapid in its headwaters among the Piedmont hills, but sluggish and marsh-bordered for the last five miles of its course. A century ago it was navigable, and trans-Atlantic shipping embarked and debarked at Bladensburg; but now it is clogged with alluvium and barely navigable above the Washington Navy Yard. Between the rivers lies a triangular amphitheater, bounded on the west by the Piedmont scarp, on the north by a terraciform upland, on the east and southeast by low bluffs carved out of Coastal plain deposits, and opening southward through the Potomac estuary. Most of this amphitheater, together with the upland borders toward the north and west, is occupied by the city.

Southwest of the city there are extensive terraces, evidently wave fashioned but deeply invaded by erosion; north of the city the upland is similarly terraced, though broad and deep ravines interrupt the continuity of the plains; and beyond the Anacostia most of the surface represents two or more wave-fashioned plains which, although deeply scored by erosion, sometimes maintain their integrity quite to the verge of the river bluffs. The Fort Myer upland, southwest of the city, is simply the scarp of a broad terrace; Kalorama Heights and Columbia Heights toward the northwest are the salients of a similar terrace; Good Hope Hill on the southeast is a remnant of another terrace; the bluff on which the National Asylum of St. Elizabeth is located is the scarp of a lower terrace of wonderful horizontality and continuity. Farther westward and northward the surface rises in less regular divides, crests, knobs and spurs; but here and there terrace remnants are found up to over 400 feet above tide, or nearly to the greatest altitudes of the region.

The terrace plains are built; the broad, low, wave-fashioned plains flooring the amphitheater are composed of the newest deposits of the region; the higher terraces carved on the walls of the amphitheater are of earlier yet late Tertiary origin. The smaller ravines as well as artificial excavations reveal the materials of the terraces in hundreds of exposures; the larger ravines as well as artificial excavations reveal the clastic formations beneath and east of the city in numberless exposures; the Potomac river and its larger tributaries are bound between steep, often precipitous, walls of the crystalline rocks. The entire region is dissected by water-ways and by a multitude of storm-cut ravines, and so the local relief is strong except toward the interiors of the broader terraces.

THE GENERAL GEOLOGY.

THE ROCKS OF THE PIEDMONT PLATEAU.

Present State of Knowledge.—Since the beginnings of American geology the prevailing crystalline character of the Piedmont terrane has been recognized, and the rocks have commonly been referred to the Archean and frequently correlated on petrographic ground with the Huronian, Laurentian and other ancient rock systems of distant parts of the country. During the last decade Dr. George H. Williams began systematic work upon Piedmont rocks in the vicinity of Baltimore ; more recently his studies have been extended westward across the entire zone along several lines in Maryland, Virginia and North Carolina. The more important results of these researches have been published by the Geological Society of America.* By means of these studies the petrographic character, structure and relations of the Piedmont rocks about the latitude of the National Capital have been made known.

The Rocks and their Relations.†—The Piedmont plateau is divisible into an eastern highly crystalline and a western semi-crystalline portion. The former consists of gneisses and holocrystalline mica schists, quartzites and marble, containing an abundance of more or less dynamically metamorphosed eruptive masses. All of these rocks have a prevailing north-northeast strike and a westerly dip. The western portion, on the other hand, is composed of partially metamorphosed sedimentary strata (sericite and chlorite schists, ottrelite schist, phyllite and limestone) and is nearly free from ancient eruptives. The strike of these rocks conforms to that of the eastern portion, but their dips are prevailingly toward the east. In spite of apparent conformity and even indications of transitions between these two portions of the Piedmont region, they are separated by a great time-break and unconformity. The easterly dips on the west and the westerly dips on the east, together with the nearly vertical strata between, produce a radiating or fan-structure, and the axis of this fan is not coincident with the contact between the crystalline and semi-crystalline portions. The thickness of either series of rocks, as indicated by their present dips, would be so vast that we must assume that the same beds are repeated over and over again by tightly compressed folds or thrusts. In the absence of all paleontologic data it is impossible to assign a definite age to either of these series. In the light of what has been discovered elsewhere, however, it is not improbable that the western and semi-crystalline areas represent the older Paleozoic horizons, metamorphosed by more intense dynamic action than has affected

* The Petrography and Structure of the Piedmont Plateau in Maryland Bull. Geol. Soc. Amer., vol. 2, 1890, pages 301-322.
† By George H. Williams.

them farther west, while the holocrystalline rocks on the east are a remnant of the pre-Cambrian continent, from which the Paleozoic sediments were derived. The apparent conformity between the two regions may be explained by supposing that the highly crystalline rocks also formed the floor upon which the now semi-crystalline schists were deposited as sediments. These older rocks, already greatly altered and folded, underwent at the time of the Appalachian uplift one more final folding, which gave them their now prevailing trend and carried the overlying Paleozoic sediments with them. This supposition is also in accord with the fact that several closed synclinals of slate and semi-crystalline schists are found pinched into the gneisses far to the east of the main contact.

The Formations of the Coastal Plain.

Present State of Knowledge.—Although geologic reconnaissance was extended over the portion of the Coastal plain lying in the middle Atlantic slope early in the present century, detailed surveys were not made until long after. So, while the composition, structure and age of the deposits were known in general terms, little was known of the precise limits of the several formations or of the geologic history recorded within them (particularly about the National Capital) until the middle of the last decade. Soon after the organization of the present Geological Survey systematic study was initiated ; within the next three years certain formations were discriminated and classified, and the methods of investigation applicable in this distinctive if not unique geologic province were developed. Subsequently detailed surveys were undertaken, under the auspices of the Geological Survey, by Mr. Nelson H. Darton. Certain formations were by him discriminated and classified, and the composition, attitude and precise areal distribution of the formations lying between the Potomac River and Chesapeake Bay (the "western shore" of Maryland) as well in much of "tide-water Virginia" were ascertained. The areal distribution of the clastic formations developed about Washington, as determined by Mr. Darton, is represented on an accompanying map ; maps of other portions of the Coastal plain are not yet published.

The surveys north and south of the Potomac-Chesapeake peninsula, and of the peninsula lying east of Chesapeake Bay (the "eastern shore" of Maryland and Delaware) are not yet completed. Accordingly, while the formations enumerated below are probably representative of the Coastal plain throughout much of the middle Atlantic slope, they are in general accurately known only in the immediate vicinity of Washington.

In the researches within the Coastal plain certain methods, developed as the work progressed, have been constantly used ; and since these methods are distinctive, and since moreover they affect materially the results of the work, they may briefly be stated :

44

(1). Reconnaissance and preliminary surveys showed that the Coastal plain deposits are commonly thin but extensive, and each composed of distinctive materials, only a part of the series being fossiliferous. Moreover the Coastal plain is vast, extending over fully 15° of latitude and 25° of longitude, and including the deposits of the greatest river of the continent, of many variously conditioned rivers of less size, and of coasts receiving little terrestrial drainage ; from which it was inferred that the distribution of organisms during past eons was affected by diverse conditions of environment, much as the fauna and flora of the present are affected. Accordingly it was deemed feasible to define the formations by composition, attitude and physical relations, and to trace formations from place to place throughout the province by means of stratigraphic continuity, independently of fossil remains, presumptively varying from place to place with the varying environmental conditions of the periods of deposition. Thus the formations discriminated in the Coastal plain are essentially physical units.

(2). As research progressed, it was found that in many cases the materials of the successive Coastal plain deposits may be traced to their sources, and that their character and distribution indicate the proximity of shores, the depth of waters, the positions and characteristics of sediment-bearing rivers, etc. Thus it was found that each formation represents a certain general relation between sea and land, the recognition of which easily explained local variations in the physical condition of the deposits ; and thus the tracing of the formations by stratigraphic continuity was facilitated and extended. So each Coastal plain formation is a physical unit, and at the same time an expression of the general physiography of the continent during the period of its deposition.

(3). As researches into the relations of land and sea during the several eons progressed, it was found that in many cases the character and distribution of deposits composing the formations indicate not only the position and size of sediment-bearing rivers, but the declivities and other conditions of those rivers, which in turn indicate the attitude, altitude and general configuration of the land surface during the period of deposition. It was also found that in many cases the land-forms themselves record geologic history definitely and intelligibly as the deposits from which history is commonly read ; and accordingly the deposition-record was in many cases supplemented by the degradation-record. So, many of the Coastal plain formations not only represent general physiographic conditions, but yield detailed records of geography and topography during the periods of deposition.

(4). As the discrimination of successive deposits of the sea and of the variously superimposed topographies of the land in the Coastal plain and Piedmont provinces progressed, it became evident that any local tract gives a record of a certain series of physical episodes, each of definite character, and that recognition of the conditions of each episode facilitates the tracing of deposits from place to place, even throughout the entire Coastal plain and far within the contiguous provinces of concurrent degradation. Thus it was found feasible not only to correlate formations with aspects of the land in each tract, but to correlate the tracts of a vast area by means of genetic identity, or by homogeny.[*] So, certain of the Coastal plain formations discriminated in the Atlantic slope represent not simple records of local physiographic conditions, but exact indices of geographic and topographic conditions extending over a considerable fraction of the continent.

For these reasons the taxonomy of the Coastal plain formations is largely independent of the paleontologic scale. Accordingly, while each formation is known to record a definite episode of continental history, its paleontologic position can seldom be indicated with accuracy in the present state of knowledge, and perhaps cannot be ascertained until researches have extended over the entire Coastal plain, and until the distribution of organisms during each episode in Coastal plain development is determined with precision.

The Formations and their Relations.—The Clastic formations found in the middle Atlantic slope, the geologic groups to which they are provisionally assigned, the thickness, attitude, and certain other characters of each, the history indicated by their physical relations, together with the approximate paleontologic position of each episode (whether of deposition or degradation), are indicated in the accompanying table ; the distribution, as determined by Mr. Darton, being shown in the accompanying map :

FORMATION.	CHARACTERS.	PALEONTOLOGIC POSITION.
Pleistocene.		
Alluvium ......Thickness unknown ; chiefly below tide ; undisturbed†.... }	Late Pleistocene and modern.
Erosion interval ; dissection of Columbia	Pleistocene.
Columbia ......Thickness 5-40 feet, altitude 150 feet ; undisturbed........	Early Pleistocene.
Erosion interval ; extensive invasion of Lafayette..............................	Pliocene (?)
Neocene.		
Lafayette ......Thickness 5 -50 feet ; altitude 500 feet ; undisturbed	Pliocene (?)
Erosion interval ; extensive planing of Chesapeake......	Miocene (?)
Chesapeake.... Thickness 10-125 feet ; tilted slightly ; fossiliferous	Miocene.
Erosion interval ; extensive planing of Pamunkey and Severn	?
Eocene.		
Pamunkey......Thickness 3-100 feet ; tilted slightly ; fossiliferous	Eocene.
Erosion interval ; extensive planing of Severn and Potomac.............	?
Cretaceous.		
Severn ....Thickness 2-25 feet ; tilted seaward ; fossiliferous	Cretaceous.
Erosion interval ; profound dissection of Potomac............................	Cretaceous.
Potomac.........Thickness 5-500 feet ; considerably tilted ; fossiliferous........	Early Cretaceous.
Long interval of extensive and profound erosion.....	Jurassic (?)

[*] American Journal of Science, third series, Vol. XL, 1890, page 36.

† Except by a late Neocene displacement which is yet in progress (c. f. 7th Ann. Rep. U. S. Geol. Survey, 1888).

There is a notable dearth of alluvium throughout the middle Atlantic slope; west of the "fall line," which is not only the common boundary of two strongly distinguished provinces but a line of modern dislocation as well, the land is rising so rapidly that the rivers, albeit rapid and generally rushing torrents, are unable to cut their channels down to baselevel; east of the "fall line" the land is sinking so rapidly that deposition in the estuaries, albeit localized and rapid, does not keep pace with the sinking.

Anterior to the vaguely limited period which may be assigned to alluvium deposition the land stood higher than now, for the antecedent formations are deeply and broadly trenched by the Potomac, the Anacostia, and other Coastal plain rivers; but whether it was the entire region or only the now sinking Coastal plain that formerly stood higher is not certainly known. It seems probable, however, that both Piedmont and Coastal provinces were elevated after Columbia deposition, that both were subsequently depressed to some extent, and that while the downward movement of the Coastal plain continues, the movement of the Piedmont plateau was long since reversed.

· The Columbia formation* commonly consists of brown loam or brick clay, grading downward into a bed of gravel or bowlders. Toward the embouchures of the larger rivers from their Piedmont gorges the loam commonly thins, and the bowlder bed thickens; in the remoter parts of the estuarine valleys the loam thickens, the bowlder bed thins, the materials become finer, and a sand bed often separates loam and gravel; farther down the estuaries the gravel bed commonly disappears, and the loam becomes interstratified and sometimes intermixed with silt. Between the rivers the deposit extends over divides up to altitudes of about 150 feet in the latitude of Washington, increasing northward and decreasing southward; and in such interstream areas the deposit is more heterogeneous than along the rivers, and contains a considerable element of materials corresponding with those of the immediate subterrane. As a whole the deposit evidently represents littoral and chiefly estuarine deposition. The materials differ from those of the modern alluvium in (1) greater dimensions of the bowlders, (2) greater coarseness of sediments in general, and, (3) less complete trituration and lixiviation of the several elements. These differences are indicative of long, cold winters, heavy snow-fall, and thick ice, but not of glaciation (in this latitude) during the Columbia period.

The Columbia formation has been traced throughout the greater part of the Coastal plain from the mouth of the Hudson to beyond the Mississippi, or over an area of more than 200,000 square miles, its thickness and composition varying from place to place with the volumes of rivers and with the character of sediments transported by them; and the altitudes of occurrences indicate submergence decreasing

---

* Defined by McGee in 1885; c. f. American Journal of Science, third series, Vol. XXXV, 1888, page 125.

from fully 400 feet in the latitude of New York to 150 feet at Washington, and perhaps 75 feet in the latitude of Cape Hatteras, thence increasing to nearly or quite 700 feet on the Savannah, diminishing next to less than 50 feet at Mobile bay, and again increasing to variable maxima farther westward and northwestward.

Traced northward the formation is found to pass under the terminal moraine and the drift sheet it fringes ; at the same time the size of bowlders and other indications of contemporaneous cold multiply, and an element of ice-ground rock flour occurs in the upper member, from which it was long inferred to represent an early episode of glaciation ; and during the present summer Salisbury has found it to pass into a premorainal drift-sheet in northern New Jersey. From the relative extent of erosion and degree of oxidation, the Columbia formation and the corresponding drift-sheet are inferred to be 5 to 50 times as old as the later glacial deposit, and a rude but useful measure of the duration of the Pleistocene is thus obtained.

During the post-Columbia period the inner gorge of the Potomac river from Washington to Great Falls was excavated. Anterior to the Columbia period the land stood so high at Washington and northward that the antecedent Lafayette formation was profoundly eroded—indeed, north of the Potomac river only isolated remnants of the Lafayette persist ; but further southward the high level diminished to such extent that the Lafayette formation maintains its continuity over wide areas. This period of erosion was long, yet not so long as to permit planation—deep and broad cañons were carved, to be subsequently converted into estuaries ; ravines were deepened and slopes steepened, and much of the Lafayette formation was degraded ; yet the interstream areas were not reduced to baselevel.

The Lafayette formation * commonly consists of well-rounded, quartzitic gravel, more or less abundantly imbedded in a matrix of red or orange-tinted loam, the gravel elements predominating in the northwesternmost exposures, and the loam predominating toward the interior of the Coastal plain. The pebbles are evidently derived from earlier members of the clastics ; the loam is derived in part from the same formation but in probably larger part from the residua of the Piedmont crystallines. The deposits differ from those of the younger Columbia formation in that the pebbles are finer, more completely water-worn, and more largely quartzitic (the Columbia alone containing bowlders and abundant pebbles of the local and sub-local Piedmont crystallines) ; and they may be discriminated from the older Potomac deposits by the smaller size and better rounding of the pebbles, and by the dearth of arkose (which is abundant in the earlier formation), as well as by a number of less striking characters.

---

* Described by Safford in 1856 [Geologic Reconnaissance of Tennessee, pp. 148, 162] and by Hilgard in 1860 [Geology and Agriculture of Mississippi, p. 3] under the name of Orange Sand ; described by McGee in 1888 [American Journal of Science, third series, vol. XXXV, p. 328] under the name Appomattox ; formally named Lafayette from original records (of 1855-56) by Hilgard in 1891 [American Geol., vol. VIII, p. 129].

48

The Lafayette formation, like the Columbia, has been recognized throughout most of the Coastal plain except in the northern portion of the middle Atlantic slope, in the Mississippi valley, and in a number of more restricted areas from which it has been degraded. Its composition varies from place to place in such manner as to indicate the local sources of material and conditions of deposition ; yet despite this local diversity it is marvellously uniform throughout the 200,000 square miles over which it has been recognized—indeed, though the youngest member of the clastic series, this formation is at the same time more extensive and more constant in aspect than any other American formation.

The Lafayette formation overlaps unconformably all the older members of the Coastal plain series in such manner as to indicate that all were extensively degraded anterior to its deposition ; yet the floor on which the formation rests is more uniform than its own upper surface, indicating that, while the antecedent erosion period was long, the land stood low, so that it was planed nearly to baselevel and seldom deeply trenched. During the post-Lafayette elevation, on the contrary, the land was deeply trenched and not planed, indicating a higher altitude than during the earlier con, but a shorter period of stream work. This record within the Coastal plain proper coincides with a geomorphic record found in the Piedmont and Appalachian zones. Throughout these zones the major and most of the minor rivers flow in broad and deep yet steep-sided gorges excavated in a baselevel plain. The Potomac gorge belonging to this category extends from Washington well toward the sources of the river ; it is within this gorge that the newer Washington-Great Falls cañon is excavated ; the same ancient gorge is admirably displayed at Great Falls, and again at the confluence of the Shenandoah at Harper's Ferry. Moreover the ancient gorges of this category are best developed in the northern part of the middle Atlantic slope, where the Lafayette formation is most extensively degraded. Now, by the concordance of history thus recorded in plain and plateau, the degradation epochs of the adjacent provinces may be correlated and the ancient gorges of the Piedmont plateau and of the Appalachian zone as well may be referred to the period of high level immediately following Lafayette deposition. While the positive evidence for this correlation is hardly conclusive, the negative evidence is more decisive—the Coastal plain deposits yield no other record of continent movement of sufficient amplitude and extent to account for this wide-spread topographic feature.

Accepting the correlation, some conception of the relative antiquity of the Columbia and Lafayette periods may be formed : In general, post-Lafayette and pre-Columbia erosion was sufficient to remove fully half of the earlier formation throughout its vast extent, and to trench it and the older formations beneath, along the present shore lines of Atlantic and Gulf, to depths ranging from 150 or 200 up to 600 or 800 feet, or to effect from 50 to 5000 times the degradation of the post-Columbia period. Again, the post-Lafayette gorges of the Piedmont and Appalachian zones exceed the post-Columbia gorges excavated by the same rivers in the crystalline rocks certainly not less than 500 times, and perhaps more than 5000 times. Moreover, if the correlation be accepted, the immense cañons of the middle Atlantic slope which, albeit more than half filled by later deposits, yet accommodate great estuaries, must be referred to the post-Lafayette high-level, and the pygmy submarine trenches of the Atlantic coast [*] must be referred chiefly, if not exclusively, to the post-Columbia high-level ; in which case the relative erosion measures are many thousands to one. It is indeed known from the steepness of wall of the Piedmont and Appalachian gorges that the excavation was effected rapidly, and hence that the land stood high above baselevel for a relatively limited period only—a period exceedingly short in comparison with the antecedent period of baseleveling ; and accordingly that the post-Lafayette high-level may not have persisted, and probably did not persist, to the beginning of the Columbia period. Yet however the several variables be evaluated, it is manifest that the pre-Columbia and post-Lafayette degradation interval must have been many times longer than the interval of degradation following the Columbia period. The relative antiquity of the Columbia and Lafayette formations thus indicated is shown graphically in the accompanying figure 1.

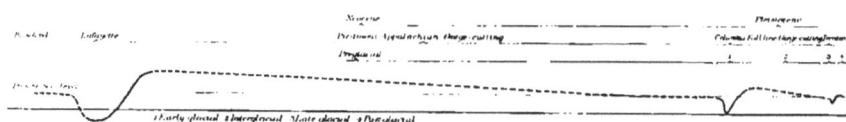

Fig. 1.

Beneath the Lafayette formation lies the Chesapeake,[†] a heavy bed of fine sands and clay, sometimes containing more or less abundant glauconite and infusorial remains and characteristic Miocene fossils. This distinctive bed is the most extensively developed member of the Coastal plain series on the " western shore " of Maryland. Although the faunas of the inferior and superior portions are somewhat diverse, the materials of the formation are essentially alike from base to summit, and the faunas intergrade in such manner that it is impracticable to divide the deposits on this ground, at least in the latitude of the National Capital. Although the formation undoubtedly extends eastward to the ocean and both northward and southward for scores or hundreds of miles, the deposits have not been actually traced much beyond Delaware Bay on the north and James river on the south.

Except as modified by the displacement coinciding with the " fall line," the newer deposits sensibly maintain the attitudes of original deposition ; but the Chesapeake and older formations are slightly deformed. This deformation, best displayed

---

[*] Recently described by Lindenkohl ; Amer. Jour. Sci., Vol. XLI, 1891, pp. 489 to 499.
[†] Defined by Darton, Bull. Geol. Soc. Am. Vol. II, 1890, p. 443.

by the surface configuration, is displayed also by the Chesapeake formation ; it consists of a slight inclination toward the fall line from an axis approximately parallel with and 4 or 5 miles distant from that boundary, together with a somewhat more decided seaward inclination beyond.

The Chesapeake formation is separated from the Lafayette above and the Pamunkey below by strong unconformities, each recording considerable degradation of the underlying formations ; but in both cases the inequalities in contact are comparatively gentle, indicating wide-spread planing rather than restricted trenching ; from which it may be inferred that the degradation period was long but that the land stood near baselevel. This deposition record of the Coastal plain has been correlated only in a general way with the degradation record of the Piedmont province ; in the latter province the extensive ancient base-level undoubtedly corresponds to several successive periods of Mesozoic and Cenozoic deposition and interruption of deposition in the Coastal plain, of which the Chesapeake period was one.

The Pamunkey formation * consists of a homogeneous sheet of sand (commonly glauconitic) and clay, with occasional calcareous layers ; and it commonly abounds in characteristic Eocene fossils. Like the Chesapeake it lies in a gentle anticlinal, its western margin inclining landward, and the great body inclining seaward.

Although it has not been actually traced on the ground beyond the limits of the "western shore" in Maryland, and "tide-water" Virginia, there are good reasons for believing that the Pamunkey formation extends throughout nearly all of the Coastal plain in the middle Atlantic slope, and probably stretches thence southward with unbroken continuity until it merges with the calcareous Eocene series of the eastern Gulf slope.

The unconformity separating the Pamunkey from subjacent formations is of the planation type, and thus tell, of a long degradation period during which the land was little elevated above baselevel. In general terms this degradation period may be correlated with the baselevel period of the plateau and the mountains ; but there are some indications that the lifting of the land was greater in the south than in the north.

The Severn formation † commonly consists of fine black, micaceous and carbonaceous sands, sometimes glauconitic, rather poorly fossiliferous, the organic remains being of characteristic Cretaceous facies. Southward from the National Capital the formation thins and soon fails ; northward it thickens and expands, undoubtedly passing into the extensive glauconitic Cretaceous beds of New Jersey. Whether the attenuation southward is due to non-deposition, to extensive degradation in this direction, or to both combined, has not yet been determined. The formation inclines seaward gently, yet more steeply than the Pamunkey ; its extension beyond the gentle anti-

* Defined by Darton ; Bull. Geol. Soc. Am., vol. 2, 1890, p. 439.
† Defined by Darton ; Bull. Geol. Soc. Am., vol. 2, 1890, p. 438.

51

clinal axis parallel with the fall line is too slight to give decisive indication of the usual landward dip of this part of the province.

The floor upon which the Severn formation rests is more uneven than its newer homologues, indicating not only extensive planation but decided trenching, and therefore may be inferred to represent long-continued degradation of land standing considerably above baselevel; yet the land record of this episode is lost in the remoteness of the period and the feebleness of the record.

The basal formation of the Coastal plain series (the Potomac *) outcrops along the "fall line" from the Delaware to the James as a heterogeneous mass of sand, clay, arkose and quartzitic or quartzic gravel. The arkose unquestionably represents the neighboring Piedmont crystallines; the quartzite is evidently derived from the extensive Paleozoic bed forming the Blue Ridge; the quartz represents the veins by which the Piedmont crystallines are frequently intersected. The more obdurate materials are not, however, confined to the Potomac formation in which they were originally deposited; they have been re-arrranged and incorporated with the Lafayette, the Columbia and probably the Chesapeake formations, and have been accumulated in modern taluses and torrential deposits. Moreover, since the advent of the white man the pebbles and cobbles have been collected for paving and guttering; and before his era they were extensively used by the aborigines for the manufacture of rude implements. Although not fossiliferous in the District of Columbia so far as known, the Potomac formation has yielded a remarkable fauna and a wonderfully rich and interesting flora. The faunal remains, collected principally between Baltimore and Washington, comprise dinosaurian bones of unique species but, according to Marsh, strong Jurassic affinities; the flora, obtained chiefly from Virginia, has been monographed by Fontaine, by whom it is regarded of Cretaceous facies and probably equivalent to the Cenomanian of Europe, though Ward deems it somewhat older.

The Potomac formation has been traced southward along the "fall line" in isolated exposures across the Carolinas and Georgia to reappear in considerable volume in Alabama, where it is designated the Tuscaloosa formation.† It has also been traced northward through Maryland and Delaware, and has been recognized in New Jersey.

The Potomac formation rests unconformably on the Piedmont crystallines, filling steep sided and narrow gorges at low levels, overspreading the moderately undulating plains at high levels. The ancient configuration revealed by this unconformity comprises an extensive Piedmont peneplain, half reduced to baselevel and afterward deeply trenched by the water-ways, much as the smoother baselevel surface of later times was trenched during the post-Lafayette high-level. The duration of the pre-Potomac degradation period was vast: At the close of the Paleozoic the eastern United States was extensively deformed, uplifted and eroded, until many thousand feet

---

* Defined by McGee in 1885; c. f. 7th Annual Report U. S. Geological Survey, 1888, p. 546.
† Bull. 43, Geol. Survey, 1888.

of the surface was carried into the sea ; then came the Newark or Triassic period of local deposition, which was followed in turn by extensive deformation, the faulting amounting probably to many thousands of feet ; and then followed comparative quietude until not only the channels of the water-ways, but the entire surface over some hundred thousand square miles was approximately baseleveled, undoubtedly by the degradation of thousands of feet of rock beds.

This sub-Potomac unconformity gives some indication of the relative position of the Potomac formation in the Mesozoic period as well as of the relative duration of the several Coastal plain periods of deposition and degradation. Let post-Columbia erosion represent unity ; then post-Lafayette degradation may be represented by 1000, and the post-Potomac and pre-Lafayette baselevel period may be represented by 100,000 ; then, using the same scale, the post-Newark and pre-Potomac erosion must be measured by something like 10,000,000, and the post-Carboniferous and pre-Newark degradation by 20,000,000 or 50,000,000. These figures are but rude approximations ; they are moreover in one sense misleading, since degradation undoubtedly proceeded much more rapidly during the earlier eons ; yet they give some conception of the relative importance of a long series of episodes in continent growth, and indicate definitively the wide separation of the Newark and Potomac periods.

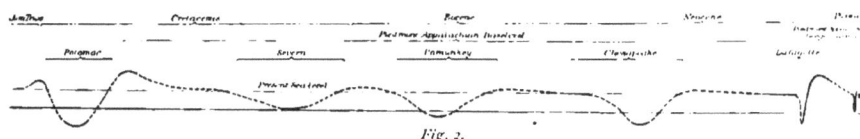

Fig. 2.

The time relations between the post-Potomac formations are represented graphically in the above figure 2. The intervals are of course only rudely approximate, yet they stand for estimates, not guesses.

THE GEOLOGY OF THE APPALACHIAN ZONE.

Present State of Knowledge.—The general features of this province were long ago made known by the classic work of the Rogers Brothers in Pennsylvania and Virginia ; but since the expansion of the field of geologic science during recent years it has been found necessary to survey in greater detail much of the area already once or twice traversed. The Federal surveys of the southern and central Appalachians were for some years in charge of Mr. G. K. Gilbert, and more recently have been carried on by Mr. Bailey Willis. One of the results of this work has been to raise questions as to the validity of the early correlation of the central Appalachian series with that of New York, except in a general way—the great groups of New York are indeed known to occur throughout the Appalachian province, yet the minor subdivisions with their distinctive faunas are found to undergo modification of such character and extent as to indicate that identity in each particular case can be determined only by more extended and detailed studies than have thus far been made. Another result has been the discrimination and delimitation of certain well-defined formations in Virginia, West Virginia, Maryland, Tennessee, North Carolina, Georgia and Alabama ; but the rocks of a considerable part of the province remain to be classified in accordance with the modern method. For the present it will suffice to say that an essentially complete American Paleozoic series of rocks is represented in the province.

The Origin and Relations of the Rocks.*—The Appalachian Paleozoic province is characterized by the occurrence of sediments deposited in the Mediterranean sea of North America, which existed during the lapse of time from the early Cambrian to the close of the Carboniferous period. It is bounded on the north and east by ancient crystalline rocks, the bases of a great mountain system, now deeply eroded, and the remains of a continent whose former extent is only to be inferred from the enormous volume of sediments it yielded to the Paleozoic sea ; and on the south and west, Mesozoic and Cenozoic deposits limit our observation of the older strata.

The history of subsidence and uplift of erosion and sedimentation may be summarized as follows :

Cambrian : The invasion of the sea, which began the known deposits of Cambrian strata along the Appalachian crystalline area, found a continent mantled in the products of rock disintegration.† These materials, easily swept away, produced a mass of fine sandstones and shales, and near the source they retained fragments of feldspar, hornblende, and other minerals, which gave rise to transition beds between the clearly crystalline and the clearly sedimentary rocks. Limestones formed where the mechanical debris was not too abundant, and the result is a complex of deposits measuring 7,000 feet and more in thickness. The uppermost member is the Potsdam, a sandstone in its typical locality, elsewhere a shale or a limestone carrying the characteristic upper Cambrian fossils.‡

Lower Silurian : This period is divided into two epochs, separated by an interval of erosion of the earlier member. The conditions of deposition continue generally unchanged from Cambrian into Silurian time, the principal result being a great thickness of chert-bearing dolomite. This formation is the most widespread, the most uniform and the most massive of all the Paleozoic series. From Massachusetts and New York to Alabama, and westward under the Mississippi valley, it is everywhere the great limestone member of the stratigraphic column. It is usually 3,000 to 4,000 feet thick. This phase of deposition was closed by an uplift, which permitted the

* By Bailey Willis.
† Pumpelly, R., "Secular Rock Disintegration, etc." Bull. Geol. Soc. Am., Vol. II.
‡ Walcott, C. D., Cambrian Faunas, Bull. 30, U. S. Geol. Survey.

formation of wave-wrought conglomerates and sea-cliff debris from the limestone along the coast line in Tennessee and in Massachusetts, and probably throughout the entire interval where detailed search has not been made. This brings us about to the close of the New York Trenton formation.

The second epoch began with the transgression of the sea, and continued until the coast line of the Cambrian ocean had been submerged. The conditions of the source of sediments were precisely like those that existed during the Cambrian, and a very similar series of conglomerates and sandstones were formed. The submergence of the land was deeper than any that preceded or followed it; sediments to a depth of 1,200 feet accumulated locally and thinned out westward to a few hundred feet. About Cincinnati they are represented by the highly fossiliferous shales and limestones of that name.

Upper Silurian : The preceding period closed with an uplift, which is possibly contemporaneous with the unconformity locally evident in the northeastern province. The first deposit of the Upper Silurian is a widespread sandstone, of peculiarly clean character, followed by the ferruginous shales of the Clinton formation, which contain the important fossil iron ores. The later history of the period is recorded in limestones, the Niagara, Salina and Helderberg, which are best represented in New York, Pennsylvania and Ohio, and thin out or disappear southward.

Devonian : In the Oriskany calcareous sandstone, followed by the Corniferous limestone in New York, we have a lithologically variable horizon, which contains fossils of both Upper Silurian and Devonian types, and marks the transition from conditions favoring the deposit of impure limestones of the Silurian to the great subsidence under the load of mud and sand deposited over New York, Pennsylvania and Virginia during the Devonian. The lowest member of this series is a highly bituminous shale, the most persistent of all Paleozoic formations except the great limestone, although in Tennessee and Alabama it is often not over 20 feet thick. In Pennsylvania it exceeds 500 feet, and in New York the formation reaches 1,200 feet. Above these dark shales follow greenish argillaceous sandstones, succeeded by red shales and sandstones. The total thickness of these mechanical deposits exceeds 8,000 feet in northern Virginia, but they thin out rapidly southward, and are not clearly recognized in Tennessee.

Carboniferous : The mechanical sediments of the Devonian are overlain by beds of limestone, which are sometimes shaly, sometimes massive and chert-bearing. Above these are the sandstones and conglomerates at the base of the coal measures, deposits of coarse materials spread over a vast area during a single epoch. Then ensued the conditions of alternating sea and marsh, which built up to a thickness of 3,000 to 4,000 feet the mass of sandy shales, shales, limestones and coal beds of the Appalachian coal field.

The Appalachian Structure.[*]—It has long been the assumption that the deformation of Paleozoic sediments in the Appalachian province took place at the close of the

* By Bailey Willis.

Carboniferous period. That certainly was the time of greatest development of folds and faults, but there is good reason to believe that there were initial disturbances as far back as the Trenton period. The forms of structure called "Appalachian," and often referred to as a single type, differ greatly in different regions. But they are all manifestations of one phase of deformation, namely, compression. A belt of strata extending along the old shore line from Canada to Alabama has been narrowed in a direction perpendicular to that shore by a reduction to five-sixths or four-fifths of its undisturbed width. This compression, which probably went on at several epochs during the Paleozoic age, raised long narrow arches with intermediate troughs (anticlines and synclines), and in some localities pressed these folds till they closed upon themselves. The force also produced movements (faults) along planes of weakness developed in the folding mass, movements which sheared across strata opposed to them in such a way as to slide older and deeply buried formations over the edges of younger deposits. Thus a geologic map of the Appalachian province usually represents many narrow parallel belts of strata in some regions, such as Pennsylvania and Virginia, winding around alternating anticlinal and synclinal axes ; in other districts, such as Tennessee, extending for scores of miles adjacent to a continuous fault line.

The history of Mesozoic and Cenozoic time is recorded in the Paleozoic province in geographic forms, in mountains, baselevel plains and river systems. What we have thus far read of this history is explained elsewhere.

THE LOCAL GEOLOGY.

CRYSTALLINE ROCKS OF WASHINGTON.[*]

General Features.—The entire area covered by the Washington atlas-sheet is composed of the crystalline rocks of the Piedmont plateau. These are, however, concealed in the eastern and southern portions of this area by the comparatively thin covering of Coastal plain deposits, from whose irregular and sinuous western edge they emerge to form the surface. Satisfactory exposures of these rocks are to be found only in the deep ravines cut by the streams (e. g. the Potomac and Rock Creek or their tributaries), since at the surface of the plateau their character has been obscured or obliterated by extensive superficial decay and by cultivation.

The older rocks of the Washington sheet belong entirely to the eastern or holocrystalline portion of the plateau province, as already described. They are for the most part granitoid gneisses of varying composition, which grade into wholly massive varieties of probably eruptive origin on the one hand, while they retain occasional evidence of clastic origin (obscure conglomeratic layers) on the other. Toward the west, as displayed along the Potomac section, which is nearly transverse to their strike, these rocks become somewhat more foliated and schistose as they approach the

[*] By George H. Williams.

boundary of the western or semi-crystalline area which passes near Great Falls in the extreme northwestern corner of the sheet. There are also much farther east occasional bands of very schistose rock (notably those seen along Broad Branch) which pass indiscriminately from one formation to another, and which owe their present character to unusually intense dynamic action.

The final period of orogenic disturbance which imparted to the entire Piedmont plateau, in common with the Appalachian system, its present structure, gave to the crystalline rocks within the Washington sheet a north-south strike. The occasional faint evidences of original bedding that have survived within this area seem now to accord closely with the foliation which has been developed in all the rocks, igneous and clastic alike, during the extreme metamorphism to which they have been subjected. This is a dip almost constantly to the west within the entire area, and growing more and more steep toward the west, in accordance with the general structure of the Piedmont plateau, as explained in a preceding section. Only in the extreme northwestern corner of the sheet, near Great Falls, do the rocks begin to incline very steeply toward the east.

* *Leading Rock Types.*—A partial examination (still in progress) of the crystalline formations within the the limits of the Washington sheet has brought to light the following easily distinguishable rock types, which are provisionally enumerated, although it is probable that further study will both modify and enlarge the list.

Granite, Granite-Gneiss and Gneiss ; Quartz–Orthoclase–Mica Rocks : This is by far the most extensively developed of all the crystalline formations of the area in question. It embraces undoubtedly eruptive granite, secondarily foliated (squeezed) granite, and typical gneiss, probably metamorphosed sediments. On account of their close lithological resemblance, decayed condition, and concealed contacts, these rocks cannot however at present be accurately subdivided on the map. Hence they are represented by a single color. Toward the east, notably along Sligo and Piney branches, these rocks are very massive, often quite devoid of any foliation, and are not infrequently filled with inclusions of other rocks in which characteristic granite contact minerals are largely developed. All this points to an eruptive origin, and these characters persist even where a secondary foliation has been developed in accord with the prevailing strike and dip. Farther westward the rocks appear more like typical gneisses, being banded, more micaceous and more schistose. Apparent beds of conglomerate have also been noticed in them along the south bank of the Potomac river, and near the Klingle Ford bridge over Rock Creek.

Diorite : Massive, dark green amphibole-biotite-granite.—These rocks present a marked contrast to the last type in their dark color. They always contain green hornblende, biotite, orthoclase and plagioclase, sometimes one and sometimes the other in excess. Quartz is also usually present and not infrequently rutile, sphene and epidote as well. Under the microscope they generally show evidence of profound

57

dynamic action. In all probability they represent ancient eruptive masses which have been subsequently greatly changed and recrystallized by earth-movements. They are most extensively developed around Georgetown and near Cabin John. In quarries at the former place clearly defined inclusions of other rocks have been noticed, which substantiate the theory of their eruptive origin.

Serpentine and Steatite : A few small lenticular areas of serpentine and soapstone occur within the area under consideration. They are usually closely associated with the more basic hornblendic rocks, and are, probably, like these of eruptive origin, although this hypothesis cannot as yet be considered as definitely proved.

Gabbro : Two small elongated exposures, presumably dikes, of trap-like rocks, which the microscope shows to be in all respects identical with the Baltimore hypersthene-gabbro,* occur near West Falls Church station, but have been as yet noticed nowhere else within the Washington region.

Broad Branch Schists : On the road leading northward from the Pierce Mill road along Broad Branch, a narrow band of thinly foliated sericitic, chloritic and siliceous schists is exposed. These rocks differ considerably in character and appearance from those about them, but still they grade imperceptibly into the granite and gneiss which lie both on their eastern and western sides. The belt, although quite narrow, has a considerable extent from north to south in the direction of its strike. Under the microscope all of these schists show evidence of the most extreme dynamic action. Their distinguishing characters (mineralogical composition, foliation, etc.) are clearly secondary ; and they may readily have been produced by an unusual amount of compression brought to bear on the normal material of the granite or gneiss. This schist belt is therefore probably the result of extraordinary pressure at the axis of a closed synclinal fold, rather than the product of metamorphism of beds originally distinct from those around them.

Siliceous Gneisses and Schists of Great Falls : The barrier at the Great Falls of the Potomac is an unusually siliceous, and therefore unusually hard, band in the gneiss. In some places this rock is so siliceous that it contains hardly anything except quartz and mica, and thus becomes a quartz schist. It exhibits throughout definite microscopic evidence of having been subjected to great pressure.

In spite of the considerable variety shown by this list, the crystalline rocks near Washington are much more uniform and monotonous than those forming the eastern part of the Piedmont plateau farther northward. This is particularly the case with the eruptives. The gabbros and gabbro-diorites, so abundant near Baltimore,† in Harford County, Maryland, and in northern Delaware.‡ are represented by only one

---

* Bull. U. S. Geol. Survey, No 28.
† Bull. U. S. Geol. Survey No. 28, by George H. Williams.
‡ Bull. U. S. Geol. Survey No. 59, by F. D. Chester.

very insignificant occurrence near Falls Church, Virginia, on the Washington sheet; while the various peridotites, pyroxenites and derived rocks [*] are altogether absent.

Granitic rocks are largely developed near Washington, and many of them preserve, both in their massive character and included fragments, fair evidence of eruptive origin. Nevertheless even these are far inferior in petrographic variety and interest to the undoubtedly intrusive granites, granite-porphyries and felsites occurring farther northward in Maryland.

CLASTIC FORMATIONS OF WASHINGTON.[†]

The General Structure.—In the vicinity of Washington the formations of the Coastal plain series are extensively displayed, and all the principal members are characteristically developed. The general structural relations are illustrated in the accompanying figure 3.

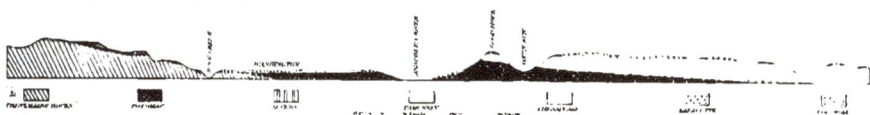

Fig. 3.—General cross-section through the central portion of the Washington atlas-sheet.

The Potomac formation lies on the steeply-sloping surface of the crystalline rocks, and is overlain by the gently eastward-dipping Severn, Pamunkey and Chesapeake formations in regular succession, upward and eastward. The Lafayette formation caps the higher plains and the Columbia formation occupies the lower terraces. The regular succession is interrupted locally in the ridge east of Washington where the Pamunkey formation overlaps the Severn for a few miles, and is in turn overlapped by Chesapeake beds which lie directly on the Potomac formation in small outliers north and west of Washington. These formations also thicken seaward, each with slightly increased rate from below upward, and their separating planes of unconformity incline gently eastward. The Lafayette formation lies across the planed surfaces of the successive formations from the crystallines westward to the Chesapeake formation eastward. The Columbia formation lies on terraces cut in the crystallines and in the Potomac, Severn, Pamunkey and Chesapeake formations from tide level to about 150 feet above.

A line of dislocation extends from northeast to southwest across the western half of the Washington atlas-sheet. The downthrow is on the eastern side and amounts to from 50 to 100 feet. This fault traverses the Lafayette and Potomac formations and the crystalline rocks, and in the divides its presence is marked by an escarpment of crystalline rocks, usually capped by Potomac and Lafayette deposits. The

[*] Am. Geol., July, 1890.
[†] By N. H. Darton.

59

date of the movement was mainly between the Lafayette and Columbia, but apparently some movement has taken place since the deposition of the Columbia.

The Columbia Formation.—The lower terraces of the Potomac valley and its larger branches and the valley of the Western Branch of the Patuxent are occupied by the Columbia formation up to altitudes varying from 80 to 145 feet. About the city or Washington the more general Columbia terrace levels are at 40 and 80 feet respectively above tide ; the Capitol being situated on the western edge of a prominent outlier of the 80-foot terrace. The formation exhibits its typical development in the District of Columbia, where it consists of two members ; a lower series of gravels and an overlying brown or buff loam. The gravels are heterogeneous in character, comprising remains of the more obdurate material of preceding formations, in large part of local origin. The loams are often quite pure, but they are frequently intermixed with sand and pebbly streaks and disseminated pebbles. Southward from Washington the Columbia terraces border the Potomac river to widths of from one to two miles, and the materials as a whole become finer. In the Anacostia valley the formation consists mainly of brown sands with pebbly streaks, but at Washington these sands merge into the loams and gravels of the typical phase. Along the northern side of the Washington–Great Falls gorge of the Potomac there is a narrow shelf at 145 above tide, which is capped at intervals by Columbia loams and gravels. The thickness of the Columbia formation about Washington averages from 20 to 30 feet.

The Lafayette Formation.—This formation occupies portions of the wide, high plains surrounding the Washington amphitheater, especially toward the south and southeast. Its materials are mainly gravels and loams. The basal and marginal beds are in larger part gravels, usually stained buff or orange, superficially, and packed tightly in stiff loams and sharp sands. The upper beds are predominantly loamy, and farther eastward loams and fine sands with gravel streaks prevail. In the outliers north and west of Washington the formation consists of gravelly red loams. The plain on which the Lafayette formation was deposited is depressed by a wide shallow basin in which is excavated the present Potomac valley below Washington. This old basin gives rise to a series of lower Lafayette terraces adjoining the Potomac valley, the principal area of which is in the vicinity of the St. Elizabeth Asylum, where its elevation is 160 to 180 feet. The Lafayette formation extends for some distance west of the " fall line " fault in a series of outliers, usually with underlying remnants of the Potomac.

The Chesapeake Formation.—This formation underlies the high plains southeast of Washington, where it is overlain by a capping of the Lafayette formation over the greater part of its area. The formation extends to the edge of the high bluffs east of Washington, and also is caught in small outliers of the higher terrace-levels at Soldiers' Home Park, and between West Washington and Tennallytown. The formation consist of very fine grained materials, mainly sands, with a variable proportion of infusorial remains and clay. In their unweathered condition the beds are usually very compact, dark gray to olive-green in color, and massively bedded. Surface outcrops consist of soft meal-like sands of light buff color. Some clay beds occur, notably locally in the eastern part of the District. Infusorial remains are nearly everywhere present, and faint casts of molluscan remains are generally abundant in the unweathered material. The outliers in the ridges about Washington consist of buff-colored, meal-like beds, lying on an irregular surface of the Potomac sands, and in turn overlain by Lafayette gravels. The thickness of the formation increases gradually eastward and is about 125 feet in the Marlborough region.

The Pamunkey Formation.—The Pamunkey formation occupies a wide area east of Washington, and is a conspicuous member of the Coastal plain series in this region. In its unweathered condition the formation is mainly a bluish or greenish-black marl, consisting of fine-grained quartz sands mixed with varying amounts of organic matter and clay, and usually containing a considerable proportion of the mineral glauconite. On weathering the glauconite is decomposed, and its iron constituent oxidizes and stains the sands to a dull red brown or snuff color. The weathered phase is general on the surface in the regions in which the formation has long been bared of overlying formations. In the streams leading out of the high plains east of Washington the unweathered marls are often exposed, and in the region to the northwestward the formation is bare for many square miles. Fossil shells frequently occur in great abundance in the marls, and there are many prolific fossil localities within a few miles of the city of Washington.

The thickness of the formation increases eastward from two to five feet in the bluff just east of Washington to over 100 feet in the Marlborough region.

The Severn Formation.—In the vicinity of Washington this formation is a thin bed of black sands, lying between the Potomac and the Pamunkey formations east of the Potomac and the Anacostia rivers. It is the attenuated southern extension of the great Cretaceous green-sand formation of New Jersey and Delaware ; but in this region it consists mainly of fine carbonaceous, more or less argillaceous, sands containing small scales of mica, but very little glauconite. It usually abounds in casts and impressions of distinct Cretaceous fossils ; and fossil shells occur in abundance at several localities. In the bluff east of Washington it is locally cut out by an overlap of the Pamunkey formation, but it comes in again toward the northeast with a thickness of 20 or 30 feet, and is occasionally exposed in streams and road-cuts throughout the eastern portion of the Washington atlas-sheet.

The Potomac Formation.—The Potomac outcrops occupy a wide area in the vicinity of Washington, especially to the southwestward. In Washington and the Potomac estuary the formation is generally hid beneath the Columbia formation, and in the plateaus toward the southwest the Lafayette formation covers it extensively. The deposits consist mainly of clays and sands of light color, commonly most irregu-

61

larly intermixed. The basal beds exposed along the western margin are mainly gray sandy arkose, with pebbles and bowlders. In Virginia the sandy arkose and arkosic sands give place eastward to gray, greenish, brown and buff sandy fissile clays. North of the Potomac they grade upward into a great series of fine quartz sands and clays, the argillaceous elements increasing in proportion eastward. Along the Baltimore and Potomac railroad, and thence eastward to the Severn formation, the clays are extensively developed, and the sands occur as locally indurated sheets and crusts, or more rarely intermixed with the clays. The formation attains a thickness of over 300 feet east of Washington, but it is eroded westward finally to a feather edge.

Post-Columbia Deposits.—The overwash deposits on slopes and along the smaller streams as well as the river muds and marshes, and the freshet deposits along the larger rivers are post-Columbia in age ; but owing to their relative unimportance they are not represented on the present edition of the geologic map. As the rivers are submerged and sinking, and the present area of submergence was preceded by erosion, alluvial deposits are mainly under water in the Washington region and consist of river muds, and the freshets are small in volume.

Artificial.—The tidal marshes adjoining the southern part of Washington have been built up above tidal level with materials obtained by excavations from the adjoining channels. This area is represented on the map as artificial.

The Geomorphology.

During recent years certain geologists have come to recognize that within certain limits earth history may be read from the land-forms developed by degradation as well as from the strata formed by concurrent deposition ; and the Coastal Plain and contiguous provinces of eastern United States are so conditioned that these lines of research may be successfully prosecuted within them.

Although the parallel mountain ranges are the most conspicuous features of the Appalachian province, the broad gently undulating intermontane plains are only less conspicuous and far more extensive ; and only less conspicuous than the intermontane plains are the narrow steep-sides gorges of all water-ways incised within the plain and sometimes notching mountains—indeed, the entire province is really an undulating baselevel plain with ranges embossed upon it, and with a series of wide-branching drainage system sharply inscribed within it.

The most conspicuous and extensive feature of the Piedmont zone is the far-stretching peneplain or undulating baselevel plain comprising the greater part of its area ; only less conspicuous are the narrow steep-sides gorges in which its water-ways flow. The Piedmont plain thus homologizes the Appalachian province with respect to classes of features ; but the embossed mountains are lacking.

Although the most conspicuous configuration of the Coastal plain is that of the present surface there are in this province a series of configurations characterizing a number of ancient surfaces, each of which is a great stratigraphic unconformity ; and the researches in this region have progressed so far that the general characters of each of these surfaces have been ascertained. The present surface is a terraced lowland, trenched by broad yet shallow estuaries and partly dissected by minor water-ways flowing in narrow steep-sided channels, produced by rapid excavation ; but portions of the lowlands are not yet invaded by the minor drainage. In general the surface is water-carved, and represents sluggish trenching along drainage lines. The next older surface (the contact surface between the Columbia and the Lafayette) is the most strongly accented of the province ; it represents a peneplain strongly and deeply trenched but nowhere planed to baselevel, save possibly in the deeper gorges far below the reach of observation. The next lower surface (the Lafayette-Chesapeake surface) is smoother than that of the present, and the configuration as well as the relations of structure to that configuration indicate widespread baselevel planation with little trenching along the water-ways. The next surface (the Chesapeake-Pamunkey) is similar, but even smoother. The Pamunkey-Severn surface in like manner is smooth and so related to the structure as to tell of extensive planation without localized vertical cutting. The Severn-Potomac surface on the other hand is decidedly rugose, and its relations to structure are such as to indicate that it represents a peneplain extensively degraded, yet chiefly along the drainage lines. The baselevel surface upon which the entire series of Coastal plain deposits rests—the sub-Potomac floor—is much like the present Piedmont surface, i. e., a rather strongly undulating peneplain, trenched by deep-cut gorges.

In addition to these general features of the three provinces there are a multitude of minor features, of which a portion have been studied and interpreted. Thus Chamberlin and Gilbert as well as White in the western part of the mountain province, and McGee in the eastern part of the same province, as well as in the plateau, have ascertained that the early Pleistocene deposits rest on the great Appalachian-Piedmont peneplain ; Willis has traced the same or a remarkably similar peneplain into the southern Appalachians in North Carolina ; Davis * has recently recognized and admirably described an ill-defined pre-Triassic and well defined pre-Cretaceous peneplain in New England and the northern Appalachians ; Emerson has incidentally developed certain features of a pre-Triassic land surface in New England ; and by these and other researches several important features in the geomorphic history of eastern United States have been elucidated. It is known that the drainage and the topographic forms resulting therefrom in the Appalachian zone were developed by orogenic movement and are therefore tectonic, they are certainly consequent in the western part of the province, and probably antecedent in the eastern part ; it is known that much of the drainage

---

* Bull. Geol. Soc. Am., Vol. II, 1891.

and configuration of the Piedmont plateau is of the subsequent type, depending upon planation and measurably reflecting rock composition, and also that another part is superimposed ; and it is known that while the principal drainage lines of the Coastal Plain are affected by relatively recent deformation the greater number of the streams and of their land-formed progeny belong to a series of autogenetic systems, repeatedly yet concordantly superimposed.

The episodes thus recognized blend as a consistent and essentially complete series of continent movements indelibly recorded in the land forms of the mountains, the plateau, and the lowland. The series begins with the faintly recorded incomplete baselevel of the pre-Triassic time ; this shadowy record is followed by the more definite one (at least in the latitude of Washington) of a long baselevel period, followed by a brief high level period during which the land first tilted seaward and then sank until the Potomac deposits were laid down ; next follows the extensive record of that long baselevel period which Davis styles "pre-Cretaceous," though it may be questioned whether this record does not merge with that of the pre-Potomac episode on the one hand and that of the long post-Cretaceous baselevel period on the other ; then follow the series of alternating episodes of sluggish deposition and indolent degradation recorded in the Severn-Pamunkey and Chesapeake formations, with their intervening unconformities—a series of episodes which may not be discriminated in the faintly inscribed record of the ancient Piedmont and Appalachian baselevel ; afterward follows the well defined episode of high level recorded in the Piedmont-Appalachian gorges and in the broad and deep trenches through which half the volume of the Lafayette formation was carried into the sea ; and then follows the inconspicuous but easily legible record of the Columbia submergence and the post-Columbia emergence—the former certified by the semi-filling of the ancient cañons, the latter by the shallow submarine channels and the pygmy "fall-line" gorges ; finally, in the northern part of the Coastal plain comes the record of submergence and subsequent lifting during the later ice invasion. This long series of generally consistent land movements is complicated in the middle Atlantic slope by the displacement, probably beginning in the Lafayette period and certainly continuing to-day ; but properly interpreted this complication only affords a check upon the accuracy of the general reading.